Evolutionary Ecology

Oxford Series in Ecology and Evolution

Edited by Paul H. Harvey and Robert M. May

The Comparative Method in Evolutionary Biology
Paul H. Harvey and Mark D. Pagel
The Causes of Molecular Evolution
John H. Gillespie
Dunnock Behaviour and Social Evolution
N. B. Davies
Natural Selection: Domains, Levels, and Challenges
George C. Williams
Behaviour and Social Evolution of Wasps: The Communal Aggregation Hypothesis
Yosiaki Itô
Life History Invariants: Some Explorations of Symmetry in Evolutionary Ecology
Eric L. Charnov
Quantitative Ecology and the Brown Trout
J. M. Elliott
Sexual Selection and the Barn Swallow
Anders Pape Møller.
Ecology and Evolution in Anoxic Worlds
Tom Fenchel and Bland J. Finlay
Anolis Lizards of the Caribbean: Ecology, Evolution, and Plate Tectonics
Jonathan Roughgarden
From Individual Behaviour to Population Ecology
William J Sutherland
Evolution of Social Insect Colonies: Sex Allocation and Kin Selection
Ross H. Crozier and Pekka Pamilo
Biological Invasions: Theory and Practice
Nanako Shigesada and Kohkichi Kawasaki
Cooperation Among Animals: An Evolutionary Perspective
Lee Alan Dugatkin
Natural Hybridization and Evolution
Michael L. Arnold
Evolution of Sibling Rivalry
Douglas Mock and Geoffrey Parker
Asymmetry, Developmental Stability, and Evolution
Anders Pape Møller and John P. Swaddle
Metapopulation Ecology
Ilkka Hanski
Dynamic State Variable Models in Ecology: Methods and Applications
Colin W. Clark and Marc Mangel
The Origin, Expansion, and Demise of Plant Species
Donald A. Levin
The Spatial and Temporal Dynamics of Host-Parasitoid Interactions
Michael P. Hassell
The Ecology of Adaptive Radiation
Dolph Schluter
Parasites and the Behaviour of Animals
Janice Moore
Evolutionary Ecology of Birds
Peter Bennett and Ian Owens
The Role of Chromosomal Change in Plant Evolution
Donald A. Levin
Living in Groups
Jens Krause and Graeme Ruxton
Stochastic Population Dynamics in Ecology and Conservation
Russell Lande, Steinar Engen and Bernt-Erik Sæther
The Structure and Dynamics of Geographic Ranges
Kevin J. Gaston
Animal Signals
John Maynard Smith and David Harper
Evolutionary Ecology: The Trinidadian Guppy
Anne E. Magurran

Evolutionary Ecology: the Trinidadian guppy

ANNE E. MAGURRAN

University of St. Andrews, Scotland

UNIVERSITY PRESS

OXFORD
UNIVERSITY PRESS

Great Clarendon Street, Oxford OX2 6DP

Oxford University Press is a department of the University of Oxford.
It furthers the University's objective of excellence in research, scholarship, and education by publishing worldwide in

Oxford New York

Auckland Cape Town Dar es Salaam Hong Kong Karachi
Kuala Lumpur Madrid Melbourne Mexico City Nairobi
New Delhi Shanghai Taipei Toronto

With offices in

Argentina Austria Brazil Chile Czech Republic France Greece
Guatemala Hungary Italy Japan Poland Portugal Singapore
South Korea Switzerland Thailand Turkey Ukraine Vietnam

Oxford is a registered trade mark of Oxford University Press
in the UK and in certain other countries

Published in the United States
by Oxford University Press Inc., New York

First published 2005

British Library Cataloguing in Publication Data

Data available

Library of Congress Cataloging in Publication Data

Data available

Typeset by Newgen Imaging Systems (P) Ltd., Chennai, India
Printed in Great Britain
on acid-free paper by
Biddles Ltd., King's Lynn

ISBN 0–19–852785–3 978–0–19–852785–5
ISBN 0–19–852786–1 (Pbk.) 978–0–19–852786–2 (Pbk.)

10 9 8 7 6 5 4 3 2 1

In memory of
Vida Macrory
1919–2004

Preface

This book seeks to capture the contribution that a small freshwater fish—the guppy, *Poecilia reticulata*, has made to evolutionary ecology. Best known as a popular aquarium species and as the subject of an Ogden Nash poem ('. . . swans have cygnets, seals have puppies, but guppies just have little guppies') the guppy has also gained prominence as a model organism in the overlapping disciplines of animal behaviour, ecology, and evolution that together form the subject matter of this monograph. The recent surge of interest in the species has been fuelled both by the recognition that the guppy is uniquely placed to answer topical questions, and by its suitability for integrated field and laboratory investigations. Indeed more than 50% of papers on guppy evolutionary ecology have been published in the last decade. However, this work rests on the solid foundation of research laid down over the last century. My book, therefore, is intended as a tribute to the early investigators who documented the patterns and posed the problems that have stimulated successive generations of researchers.

My own research on guppies began when Ben Seghers introduced me to Trinidad. I remain indebted to Ben for sharing his considerable knowledge of guppies with me. Felix Breden, Gary Carvalho, Darren Croft, Christine Dreyer, Lee Dugatkin, John Endler, Doug Fraser, Jim Gilliam, Jean-Guy Godin, Lorenz Hauser, Andrew Hendry, Anne Houde, Michael Kinnison, Astrid Kodric-Brown, Jens Krause, Kevin Laland, Robin Liley, Cock van Oosterhout, John Reynolds, Helen Rodd, Nilla Rosenqvist, David Reznick, and Paul Shaw are just some of the many researchers in the guppy community who have provided advice, encouragement, and ideas over the years. Financial support has come from the Leverhulme Trust, the Natural Environment Research Council, The Royal Society, and the University of St Andrews.

I am fortunate in having had excellent graduate students and postdocs and I thank them for their insights into fish behaviour. They include Anette Becher, Anuradha Bhat, Miguel Barbosa, Jonathan Evans, Jennifer Kelley, Siân Griffiths, Lorraine Hawkins, Phil Irving, Anna Ludlow, Kit Magellan, Jerome Masters, Iain Matthews, Alfredo Ojanguren, Charles Paxton, Lars Pettersson, Dawn Phillip, Helder Queiroz, and Stephen Russell.

The Life Sciences Department at the University of the West Indies, St Augustine, is my academic home in Trinidad. Mary Alkins-Koo, Peter Bacon, Grace Sirju-Charran, Jake Kenny, Dawn Phillip, Rajindra Mahabir, Indar Ramnarine, and Chris Starr have made me welcome and have generously provided facilities and hospitality.

Indar has been unfailingly supportive while Raj's field expertise has played a crucial role in many investigations.

Other friends and colleagues who have helped in diverse ways include John Armstrong, Jerry Coyne, Nalini Dass, Marian Dawkins, Sean Earnshaw, Jeff Graves, Peter Henderson, Felicity Huntingford, Ian Johnston, Constantino Macías-Garcia, Isobel Maynard, Manfred Milinski, Francis Morean, Martin Nowak, Geoff Parker, Linda Partridge, Gerard and Oda Ramsawak, Henry Rae, Mike Ritchie, Peter Slater, and Victoria Soo Poy.

Christine Dreyer, Jonathan Evans, Doug Fraser, Jennifer Kelley, Tino Macías-Garcia, and Indar Ramnarine commented on the draft: mistakes and omissions that remain are entirely my own responsibility.

Finally, I wish to express my gratitude to Paul Harvey and Bob May for encouraging me to write this book and for reading the manuscript, to Keith Horne for his support, and to Ian Sherman for making it happen.

Anne Magurran
St Andrews

Contents

1

Preview

The guppy, *Poecilia reticulata*, is one of the world's most widely distributed tropical fish. It is found in every continent apart from Antarctica and its range continues to be extended, both through the pet trade and as a means of controlling malarial mosquitoes. Guppies occur in some unlikely locations, such as the Moscow sewage works (Zhuikov 1993) and the River Lee in Essex in England (Maitland and Campbell 1992; but see Wheeler *et al.* 2004), where they can survive because heated effluent maintains the water temperature at tropical levels. Guppies were even sent into space aboard the USSR biosatellite Cosmos in 1987. This spectacular dispersal, much of it human assisted and not all of it beneficial to the native fish communities into which guppies are introduced (see Chapter 7), illustrates the adaptability of the species. However, it is the ability of guppies to thrive in different ecological communities and environmental conditions within their natural range of NE South America, and in Trinidad in particular, that has proved particularly fruitful in testing key evolutionary theories. These field studies and manipulations have been supported by careful laboratory experiments and have ensured the adoption of the guppy as a model organism (Amundsen 2003). The aim of this book is to explore the role that this little fish has played in shaping evolutionary ecology.

Evolutionary ecology, which I define as the interface between ecology, evolution, and behaviour (see below), is a vigorous biological discipline. Many journals are devoted to the field and papers on evolutionary ecology regularly appear in *Nature* and *Science* and dominate *Proceedings of the Royal Society: Biological Sciences*. What most investigators do not realize, however, is that pioneering research on many of the themes that interest us today was conducted using the guppy. For example, sperm competition, the subject of around 150 papers per annum in recent years, was first investigated in guppies, not by Øjvind Winge in 1937, as is sometimes assumed, but by his colleague Johannes Schmidt, two decades earlier (Schmidt 1920). Similarly, some of the earliest experiments on sexual selection were performed on guppies by Caryl and Edna Haskins (1949). The Haskinses also had the idea of transplanting guppies to sites with different predator assemblages (Magurran *et al.* 1992; Shaw *et al.* 1992). This approach was taken forward by John Endler and David Reznick and has resulted in textbook demonstrations of evolution in the wild (Futuyma 1998). Robin Liley's (1966) work on reproductive isolating mechanisms, along with Ben Seghers's (1974*a*) research into geographic variation in behaviour, illustrated the power of comparative analyses of evolution, both within and between species, and stimulated later cohorts of researchers to formulate their hypotheses in the context of the Trinidadian guppy system.

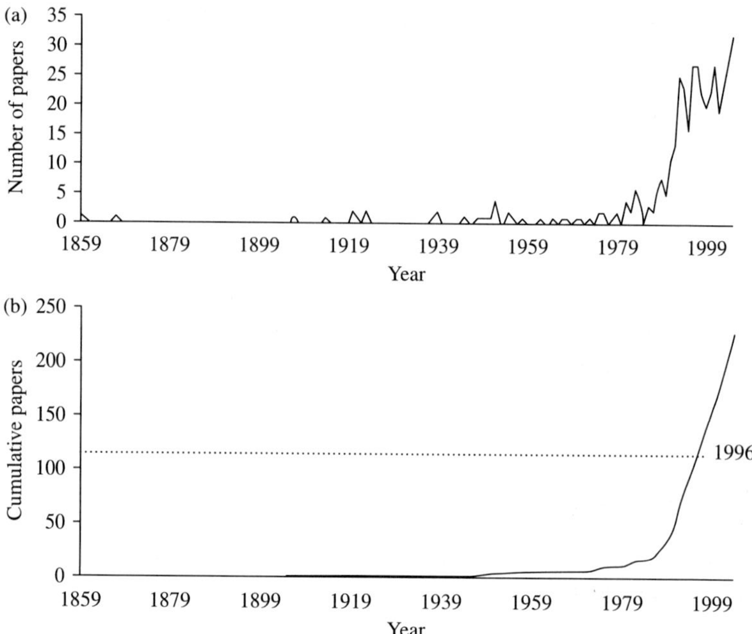

Fig. 1.1 Growth of interest in the guppy system. (a) Number of papers per annum, starting with the initial description of the species in 1859 until August 2004, on guppy evolutionary ecology. (b) Cumulative number of papers published on Trinidadian guppies—that is studies that took place in Trinidad or used fish collected in Trinidad. Fifty per cent of these have been produced since 1996 (indicated by dotted line). The reference list at the end of the book includes the papers used to construct this figure.

The popularity of the guppy in general, and the Trinidadian system in particular, as a vehicle to test ideas in evolutionary ecology is shown in Fig. 1.1. As this graph also reveals, there has been a recent upsurge of interest in this taxon with over half of all publications produced in the last decade. Chapter 2 examines the reasons for this popularity and explains why the Trinidadian guppy system is a uniquely tractable one. In essence, the accessibility of Trinidad, combined with the adaptability of the guppy to different habitats and to captivity, a short lifespan, and a wealth of background data, makes it possible to undertake the sort of investigations that are usually restricted to invertebrate models, and to complete them on a time scale that fits into the academic cycle of Ph.D. theses and research grant rounds. The opportunity to frame research questions in the context of the ecology and evolutionary history of the species increases the impact of the results. Amundsen (2003, p. 31) coined the term 'fruitflies of fish reproductive behaviour' with guppies in mind. At the same time increasing interest in the system gives rise to the concern that researchers may be adversely impacting the unique ecological experiment that fuels their research. This is a growing, but as yet little recognized problem. Chapter 7 addresses this issue.

1.1 Evolutionary ecology in the context of this book

In this book I am particularly interested in how species, in this case the guppy, adapt themselves to their environment. To put it another way I want to examine the factors that increase the mating success of individuals since it is variation at the level of the individual that determines which traits are passed on to successive generations and seals the fate of populations and species as a whole. I explore interactions with conspecifics as well as with members of different species since the biotic environment is as important as the physical environment in driving evolution. I try to identify the evolutionary responses to the selective pressures imposed by the environment. Ricklefs and Miller (1999, p. 729) define evolutionary ecology as 'the integrated science of evolution, genetics, adaptation, and ecology'. I add behaviour to their list as it is crucial in survival and reproduction.

1.2 A primer of guppy biology

In order to set the scene for the discussion in the remainder of the book I begin with a brief introduction to guppy biology.

1.2.1 Nomenclature

The guppy was first described from Venezuela as *Poecilia reticulata* by Wilhelm Peters in 1859 and independently from Barbados in 1861 by De Filippi as *Lebistes poeciloides*. A few years later R. J. Lechmere Guppy, a Trinidadian naturalist (and not a clergyman as is sometimes assumed), sent specimens of the species to the British Museum (Natural History) in London where they were named *Giradinus guppyi* by Günther (1866) in his honour. Regan (1913) recognized the confusion in the British collection and re-classified the species as *Lebistes reticulatus*. This persisted until 1963 when Rosen and Bailey restored the original name, *Poecilia reticulata*. Guppies have also been placed in the genera *Poeciliodes*, *Haridichthys*, *Acanthophacelus*, and *Heterandia* at various times and lay claim to at least 12 synonyms. Although Lechmere Guppy is no longer remembered in the scientific name of the species he is immortalized in its popular name. Guppies are also known in Trinidad as millions fish, a designation that reflects their high local abundance, and as seven colours, rainbow fish, and red tails, names that resonate with the many investigations of sexual selection conducted on the species.

1.2.2 Taxonomy and phylogeny

Guppies are poeciliids, a group of fish characterized by internal fertilization, viviparity, and the male intromittent organ, the gonopodium. Rosen and Bailey (1963) reviewed the relationships among cyprinodontiform fishes; their family Poeciliidae is still widely recognized today. Parenti (1981), however, argues that poeciliid fish more correctly belong to the sub-family Poeciliinae, which is equivalent to Rosen and Bailey's family Poeciliidae. The family (or sub-family) currently comprises 22 genera, including *Poecilia*, *Xiphophorus*, *Gambusia*, *Heterandia*, and *Belonesox*, and over 190 species (Parenti and Rauchenberger 1989). Rosen and Bailey's (1963) revision of the family Poeciliidae also placed several pre-existing genera, including *Limia*, *Lebistes*, and *Micropoecilia*, into the single genus *Poecilia*. *Poecilia* is a widely distributed and diverse genus containing 43 species (Parenti and Rauchenberger 1989). It extends from the southern United States to southern Brazil and is found in a wide range of aquatic habitats. Species in the genus vary markedly in their mating system. In some, including the guppy, female preferences for male secondary sexual characteristics, such as colour pattern, are important. In others, such as the Liberty molly, *Poecilia sphenops*, male coercion plays a much greater role in mating outcomes (Bisazza 1993)—though as Chapter 4 will reveal, post-mating mechanisms, including cryptic female choice, have the potential to influence paternity. The genus also includes the gynogenetic unisexual Amazon molly, *Poecilia formosa*.

New molecular techniques help clarify some of the relationships among the poeciliids. A mtDNA analysis which included the guppy, *Poecilia picta* and *Poecilia parae* (Breden *et al.* 1999), suggests monophyly within Rosen and Bailey's (1963) sub-genus *Lebistes*. (*P. picta* and *P. parae* were previously classified as members of the *Micropoecilia* genus—the name Haskins uses—before being placed alongside the guppy in *Lebistes* by Rosen and Bailey). The distribution of these three species is largely congruent (see below). An interesting contrast between them occurs in male coloration. All wild male guppies have different colour patterns while male *P. picta* and *P. parae* have one and three colour morphs, respectively. Rosen and Bailey's *Lebistes* also includes *Poecilia amazonica, Poecilia branneri*, and *Poecilia scalpridens*, species that occur further south, particularly in the Amazon delta in Para State, Brazil. As Breden *et al.* (1999) note, knowledge of ancestral states—and the ability to correctly identify sister taxa—is the key to understanding the evolution of contemporary traits, such as female preferences and attractive male characters.

1.2.3 Distribution

The natural range of the guppy appears to be Trinidad, Venezuela, Guyana, and Surinam and probably Tobago (Farr 1975). Guppies are also found in a few localities in Barbados (De Filippi 1861), Cuba (Barus and Wohlgemuth 1993, 1995), and Grenada (B. H. Seghers and A. E. Magurran, personal observation) but it is uncertain whether the species colonized these islands naturally or was introduced by humans. Trinidad, in contrast, is a contintental island of very recent origin and is to a large extent biogeographically part of South America (Kenny 1995 and see also Chapter 2) (see Fig. 1.2). Molecular investigations have so far shed little light on the origin of guppies in the other Caribbean islands. Fajan and Breden (1992), for example, used mtDNA sequences to deduce that Tobago guppies (from the Hillsborough River) clustered with populations in the northern and Caroni drainages in Trinidad. An investigation involving 25 allozyme loci similarly placed guppies from the Carlilse

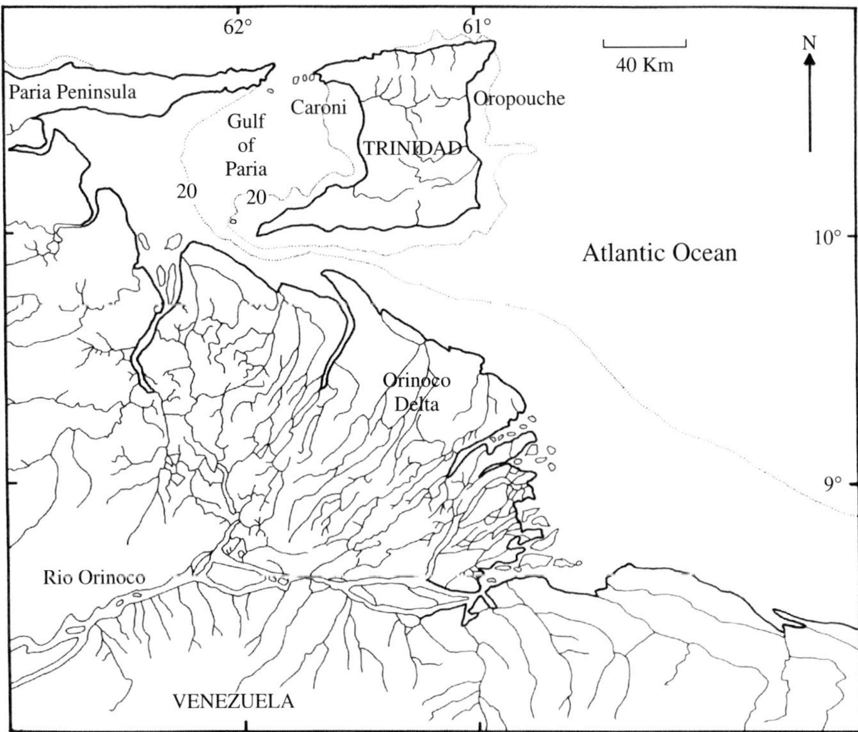

Fig. 1.2 Trinidad lies 10° north of the Equator, just off the coast of Venezuela. The locations of the Caroni and Oropouche drainages in Trinidad are shown. As the 20 (m) depth contour reveals, the water in the Gulf of Paria region is shallow. Trinidad was connected to Venezuela in the recent geological past. (See text for details).

River in Tobago firmly in the Caroni clade (P. W. Shaw, G. R. Carvalho, B. H. Seghers, and A. E. Magurran, unpublished data). Tobago guppies, along with those in Western Trinidad, are strongly differentiated from the eastern 'Oropouche' clade (Carvalho *et al.* 1991; Fajan and Breden 1992; Breden *et al.* 1999 and see further discussion in Chapter 6). (Figs. 1.3 and 1.4)

Guppies (like other members of *Lebistes*) are concentrated in the streams found along the coastal fringes of mainland South America. Guppies can even tolerate brackish water but are not typically found there. Most guppy-like fish in brackish and estuarine habitats in Trinidad and Tobago will in fact be *P. picta*. Conversely, *P. picta* may occur in freshwater but it is unusual to find the species at a distance from the sea. The relative adaptations of *P. reticulata* and *P. picta* to these different habitats, and the degree of competition between them, remain to be resolved. Mixed schools of *P. reticulata* and *P. picta* occur in a few places in Trinidad and Tobago—a point I shall return to in Chapter 6. *P. reticulata, P. picta*, and *P. parae* can be found sympatrically in Guyana (Liley 1966).

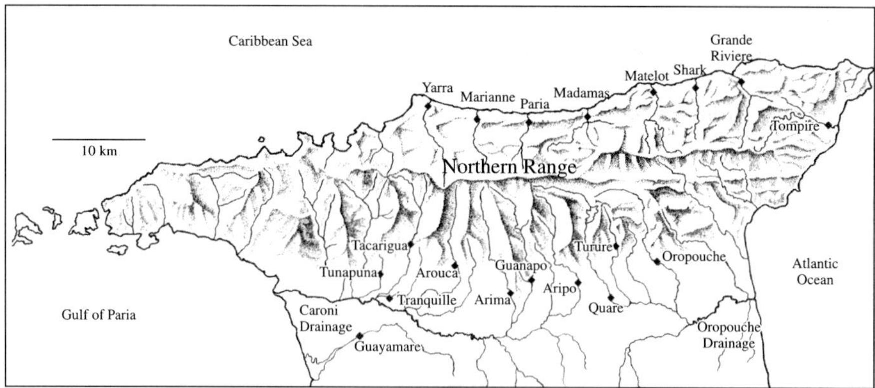

Fig. 1.3 Trinidad's Northern Range and location of key guppy populations.

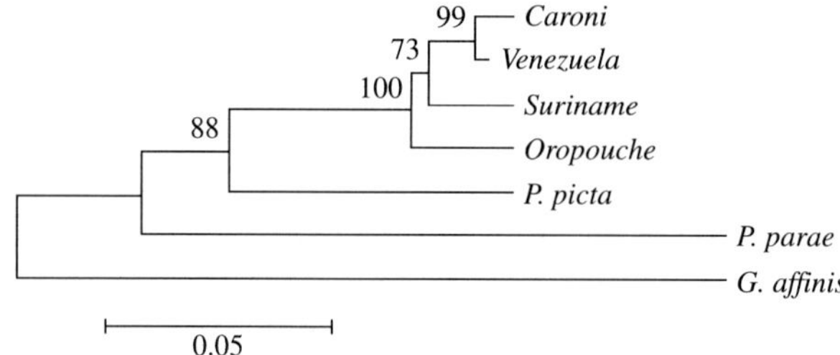

Fig. 1.4 A phylogeny constructed from Neighbour-Joining analysis of mitochondrial control region sequences (816 bp). Guppies from the Caroni (Western Trinidad) and Oropouche (eastern Trinidad) drainages are strongly differentiated. Guppies from Venezuela and Suriname, as well as *P. picta* and *P. parae* (from the sub-genus *Lebistes*) are also included. The outgroup is *Gambusia affinis*. Numbers represent bootstrap support from 1000 iterations. The scale bar represents percentage sequence divergence along a branch from the node. Sequences were obtained from Genbank. Tree prepared by S. T. Russell.

As I noted at the beginning of this chapter, guppies now occur worldwide. The first documented introduction was from Hawaii to the Phillipines in 1905 for the purpose of mosquito control (Juliano *et al.* 1989). Since guppies are not native to Hawaii it must be assumed that there were significant movements of fish prior to this date. By 1908 guppies had reached India, also for mosquito control (FAO 1997). Fishbase (www.fishbase.org) lists 53 introductions but the date of most of these is unknown.

Around 1909, Captain J. A. M. Vipan, who collected for the British Museum, shipped some live guppies to a Herr Arnold in Hamburg who is credited with their adoption as an aquarium fish. Schmidt (1920) reports acquiring his guppies, which resembled wild fish in shape and markings, from a dealer in Copenhagen in 1916. He also mentions that a short time later he procured a male guppy with large black spot on its dorsal fin from an exhibition of aquarium fish. Schmidt found, in one of the first demonstrations of Y-linked inheritance, that this marking was passed on to all the male progeny through five generations, and in all back-crosses. This male appears to have been the founder of the 'Maculatus' strain that subsequently featured in Winge's (1922*a*, *b*, 1927) classic breeding experiments and in Haskins's (1961) release experiments. Other strains, such as 'Filigran' and 'Pauper' soon followed (see Farr 1981 for a description of some of these). Meanwhile, fish breeders competed to produce lavishly ornamented males and females that bore little resemblance to their 'wild-type' antecedents. By the 1920s guppies were regularly exported from South America to Europe for research and for aquarium trade.

1.2.4 Ecology

Guppies are widespread in Trinidad and have colonized every type of freshwater body (and some brackish ones) found there but tend to be less abundant in large, deep, or fast flowing rivers than in pools or smaller streams (Kenny 1995; Magurran and Phillip 2001*a*). This distribution is mirrored in other parts of their range. Guppies can cope with a wide range of conditions and will be one of the last species to persist in severely polluted localities (Magurran and Phillip 2001*b*). This tolerance, combined with easy availability, has made the guppy the fish of choice in toxicological studies. It usually comes as a surprise to the newly arrived guppy biologist, acquainted with descriptions of typical guppy habitats from the literature, to find a flourishing population in the foetid drain that runs under the Priority Bus Route, just outside the campus of the University of the West Indies. This is not to say, of course, that guppies are impervious to poor water quality. A growing list of studies is, for example, showing that the presence of endocrine disrupters has detrimental consequences for sperm production and reproductive behaviour (Bayley *et al.* 2002, 2003). Chapter 7 will discuss threats to guppy populations further.

As Haskins *et al.* (1961) first observed, guppies are also found in a range of fish assemblages in Trinidad and show adaptations that are correlated with the intensity of predation. It is this variation that has fuelled much research and will be a recurrent theme in this book. Guppies in predator-rich assemblages, which usually, but not invariably, means they coexist with the pike cichlid *Crencichla alta*, differ in morphology, life history and behaviour from those found in less dangerous habitats. The density of guppies in the wild varies greatly over space and time but can be as high as 75 adults m^{-2} (see Fig. 2.13). Sex ratios also fluctuate dramatically and an excess of females is not uncommon (Seghers 1973; Pettersson *et al.* 2004). Sex ratios at birth are, however, not significantly different from unity (Haskins *et al.* 1961; Pettersson *et al.* 2004). Chapter 2 examines the ecology of guppies in greater depth.

1.2.5 Reproduction

All poeciliids have internal fertilization (Wourms 1981). Sperms are produced in bundles, termed spermatozeugmata, and transferred to females using the gonopodium, a modified anal fin. The structure of the gonopodium, which often sports hooks and claws, varies markedly from species to species and is an important character in species identification (Rosen and Bailey 1963). Female poeciliids can store sperm in the folds of their ovaries and gonoducts (Constanz 1989). In guppies, for example, stored sperm can continue to fertilize ova for up to 8 months (Winge 1937). Recently inseminated sperm will, however, secure most fertilizations (Schmidt 1920; Constanz 1984). Female guppies mate multiply (Winge 1937; Evans and Magurran 2000) and the median number of sires per brood is two (Becher and Magurran 2004). Within a given brood cycle the last male to mate is likely to father most offspring (Hildemann and Wagner 1954; Evans and Magurran 2001). Male guppies have two means of securing copulations. They may either display in an attempt to gain a consensual mating with a receptive female or they may engage in gonopodial thrusting—sneaky mating—and try to inseminate uncooperative females (Baerends *et al.* 1955). Females are most receptive to male courtship either as virgins or for 1 or 2 days following parturition (Liley 1966). A reproductive cycle typically lasts about 25–30 days (Haskins *et al.* 1961; Houde 1997) although there can be considerable variation in this (see, for example, Evans and Magurran 2000), particularly in females that have mated for the first time. Reproduction continues through the year (Alkins-Koo 2000) although there are some seasonal differences in reproductive investment (Reznick 1989). Like most other poeciliids guppies are livebearers (in the one exception to this rule, *Tomeurus gracilis*, eggs are laid on leaves following internal fertilization (Keith *et al.* 2000)). Guppies are often referred to as an *ovoviviparous* species (Turner 1947), meaning that the embryos are nourished by the yolk that the female deposits in the egg prior to fertilization. Constanz (1989) prefers the term *lecithotrophic*. This contrasts with viviparous or matrotrophic species in which the mother continues to nourish the egg after fertilization (Constanz 1989). Poeciliids fall along a lecithotrophic–matrotrophic continuum, with guppies, whose embryos lose 25% of dry weight during gestation (Thibault and Schultz 1978; Wourms 1981) being placed firmly at the lecithotrophic end. Superfetation describes the condition where several broods, at different stages, are carried simultaneously (Wourms 1981; Constanz 1989). Poeciliids can be either superfetating or non-superfetating. Guppies are an example of the latter. All non-superfetating poeciliids are classified as lecithotrophic while all but one superfetating species are matrotrophic (Reznick and Miles 1989). The reproductive behaviour of guppies is described in more detail in Chapter 4.

1.2.6 Life-history patterns

By the time guppies are born, they are well developed and are capable of independent existence. There is no further parental care. Indeed, baby guppies school from birth and can perform an array of anti-predator tactics (Magurran and Seghers 1990*b*)—a necessary skill given their vulnerability to cannibalism and predation. Brood size is extremely variable and ranges from a single offspring to a hundred or more. Around half of the variance among females in litter size can be attributed to female body size (Reznick and Endler 1982; Reznick and Bryga 1987; Travis 1989). There are also consistent differences among populations, with fish from high predation localities producing more, but smaller, offspring, than size-matched females from low-risk populations. For example, the expected number of offspring produced by a standardized (average-sized) female from a typical '*Crenicichla*' (see p. 15) population is 6.4. This contrasts with 2.8 for an equivalent female from a typical '*Rivulus*' population (Reznick and Endler 1982, table 2). Larger wild females, particularly those from high-predation populations, may give birth to 30 or more babies at a time (see Chapter 5). It is only in domestic strains, bred for large body size, that broods of a few dozen and above are consistently observed.

Female guppies first produce offspring at 10–20 weeks of age and there are around 2–3 generations per year in the wild. Males can mature in 7 weeks or less (Reznick *et al.* 2001*a*). There is considerable variation among localities in size at maturation, not all of it related to predation risk. For example, in the Carlisle and Quarahoon drainage in SW Trinidad, females and males mature at 11 and 9 mm standard length (SL), respectively (Alkins-Koo 2000). The Carlisle and Quarahoon system supports a diverse fauna of both predators and prey that includes species, such as the hatchet fish *Gastropelecus sternicla*, that occurs in only a handful of rivers in Trinidad. In the Northern Range, by comparison, the median standard lengths of females at the time of first reproduction are 15 mm in *Crenicichla* sites, and 18 mm in *Rivulus* sites (Reznick and Endler 1982). The equivalent figures for males, which in contrast to females virtually stop growing at maturity, are 15 and 16 mm. Females that manage to evade predators and parasites will continue to reproduce until 20–34 months of age (Reznick *et al.* 2001*a*). There does not appear to be any 'menopause' or prolonged post-reproductive period (Reznick *et al.* 2001*a*) though elderly and non-reproductive females can be observed in lab stocks. Occasionally females develop male colour patterns and a gonopodium in later life, and may even begin courting. I have never, however, observed any of these 'transgendered' fish siring offspring, despite placing several with virgin females. Male courting vigour may decline in older fish. Guppies have proved a rewarding model for testing theories of aging and senescence (Reznick *et al.* 2001*a*, 2004) and reveal how rapidly 'contemporary evolution' (Reznick and Ghalambor 2001; Stockwell *et al.* 2003) can occur in the wild.

1.3 Overview of the book

The primary aim of this book is to use a historical sweep to illustrate how this uniquely tractable system has raised key questions in evolutionary ecology and supplied many of the answers. In doing so it will reveal how durable good data are—despite shifting emphases in the subject. Indeed, one of the things that has struck me most forcibly while writing the book is how well careful observations, whether they be Winge's demonstration of sex linkage in male colour genes or Liley's account of guppy mating tactics, have stood the test of time. It is fascinating to discover how many of the issues that intrigue us today were mulled over by the pioneers in the field.

My second goal is to provide an overview of the Trinidadian guppy system and to use this to explore the evolutionary consequences of ecological processes. Geographic variation in predation risk is the most famous of these and the attendant diversification in colour pattern and behaviour a well-cited example of evolution in action. But there are interesting complexities and subtleties that are sometimes overlooked and the book provides an opportunity to draw attention to these. Throughout I will endeavour to identify unresolved questions and mention a few of the many topics that deserve further study. Although the focus is on the Trinidadian guppy, parallel problems and advances in other study systems will also be highlighted.

The book opens (Chapter 2) with an investigation of the ecology of the guppy in Trinidad. In it I discuss the types of habitat in which guppies are found, as well as the factors, particularly predation and productivity, that shape their behaviour and morphology and life-history traits. Although much of the focus in the literature is on variation in predation risk, the manner in which risk varies, over both space and time, is rather poorly understood. Moreover, risk tends to covary with productivity, making it more difficult than is often assumed to disentangle cause and effect in evolution in the wild. Chapter 3 examines the direct consequences of predation risk, primarily in the behavioural adaptations to risk. In Chapter 4 I turn to reproductive biology and emphasize some recent and very exciting work on post-copulatory dynamics. It is pleasing that the species that provided the first insights into sperm competition continues to inform our understanding of the field. Guppy populations differ not only in their mating tactics but also in how they make their investment in reproduction. Life-history traits, such as age and size at maturity, number and size of offspring, and senescence vary markedly between populations. Chapter 5 considers these issues along with the consequences of key life-history decisions for mating success in both males and females. Although guppies have become the best-known example of population differentiation in the wild it has been rather perplexing to realize that rapid evolution in morphology and behaviour has not been accompanied by the equally rapid emergence of reproductive isolating mechanisms. Chapter 6 assesses the reasons for this but also presents recent work that has uncovered isolation between the Caroni and Oropouche guppy clades in Trinidad and between guppies and their sister taxon, the Cumaná guppy, in Venezuela. These cases provide complementary insights into the evolution of reproductive isolation in a promiscuous mating system. Guppy populations can shed light on how new populations are formed; they are also at risk through a combination of anthropogenic influences. In Chapter 7, I examine how pollution, habitat loss, and exotic introductions impact natural populations of guppies in Trinidad and other places. I further ask whether we as scientists amplify these problems and what we might do to ameliorate them. Finally, Chapter 8 will draw together the main themes of the book before reviewing the legacy of the pioneer guppy researchers. I believe that the guppy system will continue to offer unrivalled opportunities to test theories in evolutionary biology, and attempt to identify what some of those might be. And I end with the plea that this irreplaceable resource be safeguarded.

2

Ecology of the guppy in Trinidad

On first inspection, the Trinidadian guppy system seems to have textbook simplicity. Nonetheless, as this chapter will reveal, there are significant complexities that make the investigation of the evolutionary ecology of the guppy both more interesting and more challenging that it initially appears.

As its rich fauna and flora testifies, Trinidad is a continental island that once formed part of mainland South America. During the last Ice Age sea levels were between 100 m and 130 m lower than at present (Kenny 1989) and the land mass took a very different form to the one we are now familiar with (Fig. 2.1). Indeed, evidence from paleo-corals in the Gulf of Paria suggests that final separation between the island of Trinidad and Venezuela may have occurred as recently as 1500 years ago

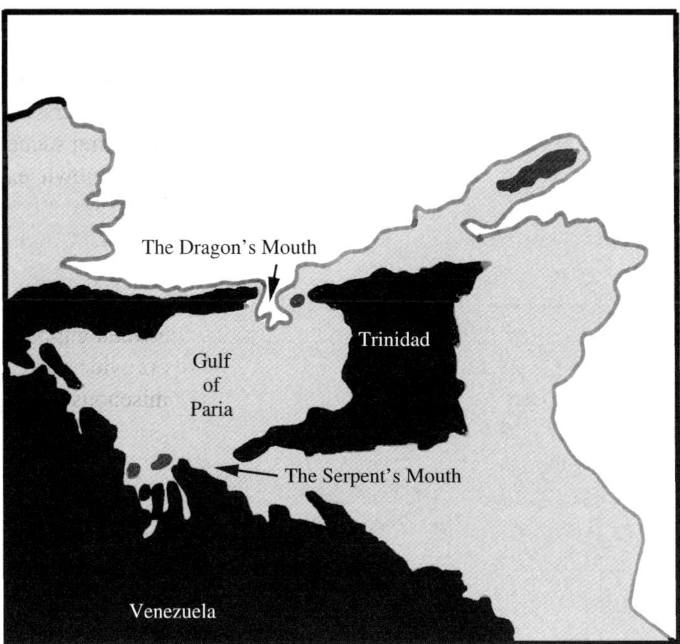

Fig. 2.1 The shoreline of the region during the early Pleistocene (1.8 million years ago). The first ingress of water into the Gulf of Paria appears to have been in the region of the Dragon's mouth with the land bridge at the Serpent's Mouth the last to be breached. After Commeau (1991).

(Kenny 1989). It seems likely that the land bridge connecting the southwest (Cedros) peninsula of Trinidad and Venezuela was the last to be breached. Prior to this the Caroni and other westerly flowing rivers in Trinidad, along with the easterly flowing Rio San Juan and Rio Manamo in Venezuela, probably drained into a deep bay in the northern part of the Gulf of Paria (Kenny 1989). This also means that the Orinoco discharge would have been deflected to the east of Trinidad and that at one time the Oropouche system may have been a tributary of the Orinoco. These historical events have left a genetic signature on guppy populations in Trinidad and provide important opportunities to unravel the evolution of reproductive isolation (see Chapter 6).

The previous chapter pointed out that one local name for *Poecilia reticulata* in Trinidad is 'millions', in part because the species is so abundant, but also because it is so widespread. Dawn Phillip and I (Magurran and Phillip 2001*a*) confirmed that guppies are the most widely distributed freshwater fish in Trinidad (see also Price 1955; see also Kenny 1995). We conducted a stratified survey of river systems and found *P. reticulata* in 80% of the 80 sites we examined (Fig. 2.2). It also ranked as the most abundant species overall in terms of number of individuals (though not in terms of biomass: here guppies were only the 14th most abundant species out of an assemblage of 41 species). Guppies occur in all freshwater environments ranging from clear, oligotrophic mountain streams to turbid, lowland rivers. The species thrives in small drainage ditches as well as in large water bodies and can even tolerate polluted conditions. Guppies are, however, absent in some remote northern streams, such as the Matelot River. These rivers drain the northern slopes of the Northern Range into the Caribbean Sea and are isolated from other drainage systems. While it is known that some populations can go extinct (see Chapter 7)—, for example, some

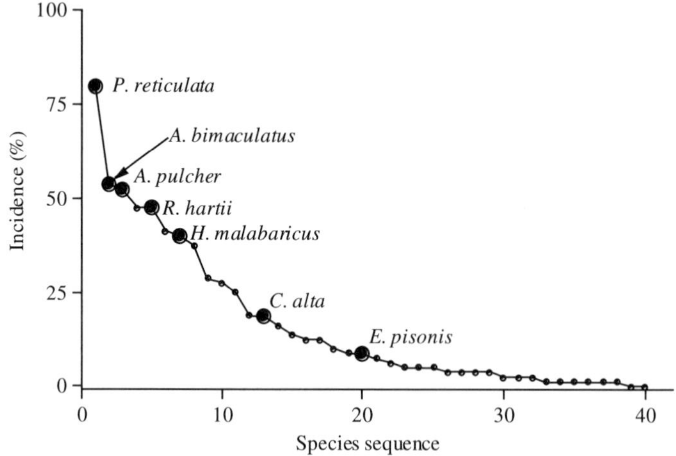

Fig. 2.2 Incidence (as percentage of sites found) of the 40 freshwater fish species in the Trinidadian assemblage. Species are ranked from most to least abundant. The identities of guppies and some potential predators are shown. After Figure 2 in Magurran and Phillip (2001).

artificial introductions fail (D. N. Reznick, personal communication)—the most likely explanation for the absence of guppies from these streams is that they were never naturally colonized.

2.1 What makes Trinidad special?

As Chapter 1 indicated, Trinidad has a number of features that make the guppy system irresistible to evolutionary ecologists. It is a geographically compact island measuring approximately 80 km and 50 km on its N–S and E–W axes, respectively. English is the official language and there are reliable airline services to Europe and North America. The New York Zoological Society set up a research station at Simla, under William Beebe, in the Northern Range in 1949. This station is still functional and is now managed by the Asa Wright Nature Centre; several other establishments also currently provide research facilities. The logistical aspects of research are thus well catered for. In addition, there is a wealth of historical and background information on the guppy system. Lechmere Guppy's pioneering account of the freshwater fauna of Trinidad (see Regan 1906) ushered in a long tradition of investigation. To date there are over 250 papers and monographs that deal directly or indirectly with Trinidadian guppies and guppy communities. The Imperial College of Tropical Agriculture, which later became the University of the West Indies, St Augustine, provided (and still provides) an important reservoir of local expertise. However, it was the ability of the Trinidadian guppy system to answer key questions in evolution, that ensured its adoption as a model system. When evolutionary ecologists began to quantify natural selection in the wild the opportunities provided by Trinidadian guppy populations, particularly those found in the Northern Range (Fig. 1.3), became apparent. Here was a vertebrate species, with a short generation time and marked—and meaningful—population differences. The particular configuration of the river systems in Trinidad and distribution of fish species, including guppies, among them, has resulted in pronounced, rapid, and interpretable population differentiation in a range of adaptive traits over a small geographical scale. The level of documentation, combined with the tractable nature of the system is such that it is possible to visit Trinidad for the first time and complete a publishable study within a few weeks. But there are downsides to this accessibility too—as Chapter 7 will reveal.

2.2 Predators

Caryl Haskins's comment, that the rivers of the Northern Range represent a natural 'laboratory' (Haskins *et al.* 1961, p. 333), was the catalyst that attracted researchers to Trinidad and is the premise on which many current investigations still rest. The southern slopes of the Northern Range are drained by a series of parallel rivers, each of which contains guppies throughout most of its course. Moreover, the geology of the region is such that many of these rivers are partitioned by barrier waterfalls of sufficient magnitude to have prevented the upstream colonization of the larger fish species that are potential predators of guppies. For example, the Lower Aripo River (below the barrier waterfall) has a rich fish assemblage that includes guppies, *Rivulus hartii* (Hart's rivulus or jumping guabine), *Crenicichla alta* (pike cichlid or matawal), *Hoplias malabaricus* (guabine or wolf fish), *Aequidens pulcher* (blue acara), *Astyanax bimaculatus* (two-spot sardine), *Hemibrycon taeniurus* (mountain sardine), *Synbranchus marmoratus* (swamp eel or zangie), and *Awaous taiasica* (sand fish). Upstream, in the Upper Aripo and Naranjo tributaries, the fish fauna consists of guppies, *R. hartii, A. pulcher* (in places), and probably *S. marmoratus*. There is thus a marked contrast in community structure—and predation risk—over a short distance (Fig. 2.3). Other factors, such as flow regime, water depth, and clarity, and

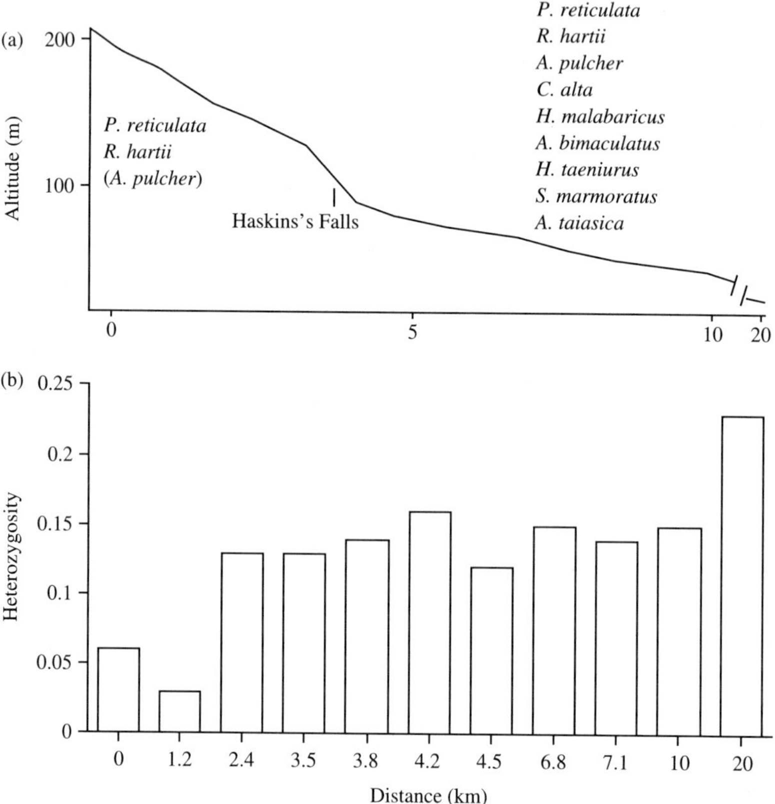

Fig. 2.3 The Aripo River. (a) The river profile. Haskins's Falls mark the boundary between the upstream and downstream low- and high-predation assemblages. Key members of these assemblages are listed. Distance from river source is indicated. 0 km marks the locality at which the first guppies are present. (b) Mean heterozygosity of guppies at various points along the river. Heterozygosity is derived from allozyme data. Distance is measured in relation to the uppermost guppy population (as in (a)). These data are taken from Shaw *et al.* 1994.

substratum, are usually broadly similar immediately above and below the barrier. These contrasts are replicated in most rivers and (although Haskins was unaware of it) in two guppy clades (that is the fish found in the Oropouche and Caroni drainages—see Chapter 6). There are also barrier waterfalls that produce parallel discontinuities in fish assemblages in several of the rivers on the northern slopes of the Northern Range. Here guppy predators are drawn from the Antillean fauna, and include fish, such as the piscivorous goby *Eleotris pisionis*, rather than from the South American fauna that dominates most of Trinidad (Reznick and Bryga 1996; Reznick *et al.* 1996*c*). The attraction of the system is obvious. There are clear differences in a major ecological factor—predation risk. Replicated analyses are possible. In addition, the system lends itself to independent tests since both the evolutionary history of the guppy populations (Caroni versus Oropouche rivers) and the composition of the predator communities (northern versus southern drainages) can be controlled.

The iconic guppy predator is the pike cichlid *C. alta*.[1] *C. alta* occurs in many downstream and lowland habitats. It was first linked to guppy evolution by Caryl Haskins (Haskins *et al.* 1961). Subsequent research has confirmed that the pike cichlid is indeed an effective predator of guppies and an agent of evolutionary change. In his classic transplant investigation Endler (1980) found that guppies released from predation by pike cichlids (and other predatory species) rapidly evolved brighter colour patterns. Many guppy biologists (myself included) have thus adopted the shorthand '*Crenicichla* site' to identify localities where predation risk is high. Conversely, a '*Rivulus* site' is indicative of low predation risk.

This classification is convenient but inevitably over-simplifies the true picture. In fact risk is extremely difficult to quantify as it depends not only on the presence of predator species but also on their relative and absolute abundance, as well as on the density of the guppy population, the structure of the microhabitat, and the availability of other prey species. In addition different ontogentic stages—newborn, juvenile, and adult—may experience different degrees of risk. Moreover risk will vary over time in line with changes in predator and prey populations. Below I summarize existing information on guppy predators.

2.2.1 Fish predators

Haskins *et al.* (1961) listed *A. bimaculatus*, *A. pulcher*, *C. alta*, and *H. malabaricus* (see Table 2.1) as the more serious fish predators of guppies in the Arima and Aripo Rivers. As noted previously, the upstream distribution of these fish is truncated by the barrier waterfalls. *R. hartii*, a 'less severe' predator, was observed by Haskins to be a nearly ubiquitous inhabitant of freshwater on the island (a conclusion that Magurran and Phillip (2001*a*) also reached following their survey of Trinidadian rivers). The Trinidadian name for *Rivulus*, 'jumping guabine' testifies to its ability

[1] Coleman and Kutty (2001) argue that the predator of guppies in Trinidad is *Crenicichla frenata* not *C. alta*.

Table 2.1 Fish species found in Trinidad—and their feeding modes

FAMILY and species (after Kenny 1995)	Feeding modes by family (after Kenny 1995)	Diet and trophic level (from Fishbase (www.fishbase.org))
ERYTHRINIDAE	Carnivorous, macrophagous, and generally ichthyophagous	
Erythrinus erythrinus		Nekton, zoobenthos (3.77±0.60)
Hoplerythrinus unitaeniatus		Nekton, zoobenthos (3.83±0.62)
Hoplias malabaricus		Nekton, zoobenthos (4.02±0.65)
CURIMATIDAE	Omnivorous, microphagous select deposit feeder	
Steindachnerina argentea		—
GASTEROPELECIDAE	Carnivorous, entomophagous, and surface feeding	
Gasteropelecus sternicla		Zoobenthos (3.28±0.41)
ANASTOMIDAE	Generally ominvorous and macrophagous, surface and mid-water feeding	
Leporinus frederici (1,2)		Zoobenthos (2.59±0.27)
CHARACIDAE	Generally omnivorous and macrophagous, surface, mid-water, or bottom feeding	
Astyanax bimaculatus		Detritus, nekton, plants, zoobenthos, zooplankton (2.88±0.31)
Brycon siebenthalae		—
Corynopoma riisei		—
Hemibrycon taeniurus		
Hemibrycon ocellifer (3)		—
Hemigrammus unilineatus		—
Megalamphodus axelrodi		—
Moenkhausia bondi		—
Odontostilbe pulcher		—
Roeboides dayi		Zoobenthos, zooplankton (3.47±0.44)
Triportheus elongatus (1,2)		—
LEBIASINIDAE	Generally omnivorous, surface and mid-water feeding	
Copella arnoldi (3)		Zoobenthos (3.26±0.4)
Nannostomus unifasciatus (2,3)		Zoobenthos (3.27±0.4)
Pyrrhulina laeta (3)		Zoobenthos (3.26±0.4)
GYMNOTIDAE	Carnivorous and generally icthyophagous, bottom and mid-water, with some surface feeding	
Gymnotus carapo		Detritus, nekton, plants, zoobenthos, zooplankton (3.56±0.56)

Table 2.1 *(Continued)*

FAMILY and species (after Kenny 1995)	Feeding modes by family (after Kenny 1995)	Diet and trophic level (from Fishbase (www.fishbase.org))
CALLICHTHYIDAE	Omnivorous, microphagous and deposit feeding, at or near bottom	
Callichthys callichthys		Nekton, plants, zoobenthos (3.28±0.47)
Corydoras aeneus		Plants, zoobenthos (2.96±0.34)
Corydoras mclanistius (2,3)		Plants, zoobenthos (2.84±0.32)
Hoplosternum littorale		Detritus, plants, nekton, zooplankton, zoobenthos (3.09±0.36)
LORICARIIDAE	Omnivorous, microphagous, and grazing	
Ancistrus cirrhosus		Plants (2±0.0)
Hypostomus robinii		Plants (2±0.0)
PIMELODIDAE	Carnivorous, scavenging, macrophagous, and bottom feeding	
Rhamdia quelen		Nekton, zoobenthos (3.56±0.49)
AUCHENIPTERIDAE		
Pseudauchenipterus nodosus		Detritus (2±0.0)
ANGUILLIDAE		
Anguilla rostrata		Nekton, zoobenthos (3.62±0.54)
SYNBRANCHIDAE	Carnivorous, scavenging, bottom feeder	
Synbranchus marmoratus		Zoobenthos (3.2±0.4)
POECILIIDAE	Omnivorous, surface, mid-water, and bottom feeder	
Poecilia picta (4)		—
Poecilia reticulata		Zoobenthos (3.2±0.40)
Poecilia sphenops (3)		Detritus, zoobenthos (2.95±0.34)
Poecilia vivipara (4)		Zoobenthos (3.2±0.40)
RIVULIDAE	Generally, carnivorous, surface and bottom-feeder	
Rivulus hartii		Nekton, plants, zoobenthos (3.02±0.45)
MUGILIDAE	Omnivorous, surface, mid-water, and bottom feeder	
Agonostomus monticola		—
CICHLIDAE	Generally carnivorous, consuming fish and invertebrates	

Table 2.1 (*Continued*)

FAMILY and species (after Kenny 1995)	Feeding modes by family (after Kenny 1995)	Diet and trophic level (from Fishbase (www.fishbase.org))
Aequidens maronii (1,2)		Zoobenthos (3.25±0.40)
Aequidens pucher		Zoobenthos (3.27±0.40)
Cichlasoma taenia		—
Crenicichla alta		Nekton, zoobenthos (3.56±0.51)
Oreochromis mossambicus (3)		Detritus, nekton, plants, zoobenthos, zooplankton (2.71±0.32)
ELEOTRIDAE	Carnivorous, macrophagous, and ichthyophagous; bottom feeder	
Dormitator maculatus (4)		—
Eleotris pisonis		Nekton, zoobenthos (3.72±0.61)
Gobiomorus dormitor		Zoobenthos (3.72±0.59)
GOBIIDAE	Omnivorous and microphagous, bottom feeder	
Awaous taiasica		Detritus, plants, zoobenthos (2.78±0.31)
Sicydium punctatum		—
NANDIDAE		
Polycentrus schomburgkii		Nekton, zoobenthos (3.61±0.53)
GOBIESOCIDAE	Probably omnivorous and microphagous	
Gobiesox nudus		Nekton, zoobenthos (3.55±0.50)
1 = natural colonist; 2 =extinct; 3 = introduced; 4 = mainly brackish.	Kenny (1995) prefers the term ichthyophagous to piscivore and distinguishes between macro- and micro-phagy.	Nekton signifies a diet composed of bony fish.

Notes: The list of fish species is taken from Kenny (1995). Feeding modes, by family, are shown (following Kenny 1995). The diet and trophic level (where available) of species is derived from Fishbase (www.fishbase.org). Fishbase estimates trophic level from recorded food items using a randomized resampling routine. The mean value and standard error of the estimated trophic level are included. Further information on feeding modes and diet is provided at the foot of the table.

to ascend barriers. It may also move over land during the wet season (Regan 1906; Haskins *et al.* 1961). For this reason *Rivulus* is often the only fish in isolated pools at the heads of rivers. Other fish that have since been added to this 'predator assemblage' include *E. pisonis* and *Gobiomoris dormitor*, species found in the lower sections of streams in northern drainages in Trinidad (Reznick and Bryga 1996).

Despite the intense interest in the evolutionary ecology of the guppy in Trinidad information of the diets and feeding preferences of putative predators is remarkably slim. Kenny (1995, p. 65) cautions: 'Unfortunately, as the main preoccupation is with predators of guppies and the guppies responses, it is sometimes forgotten that few of these predators feed exclusively on guppies'. I therefore first evaluate the potential of fish in the Trinidad assemblage to act as guppy predators and then review studies that have set out to measure their impact on guppy populations.

Table 2.1 lists the species that comprise the freshwater fish fauna of Trinidad and summarizes information on their feeding behaviour. The table also gives the trophic level of these species. To set this information in context, a fish with a trophic level of 2.0 feeds almost exclusively on plant material whereas a fish with a trophic level of around 4.0 is predominately a carnivore. Large confidence limits attest to a variable diet. Since the predatory capacity of a fish—from the guppy's eye view—is some function of its size, Fig. 2.4 plots trophic level in relation to maximum body length. All other things being equal species located towards the top right hand corner of the graph have greater potential as guppy predators. However, some of these species

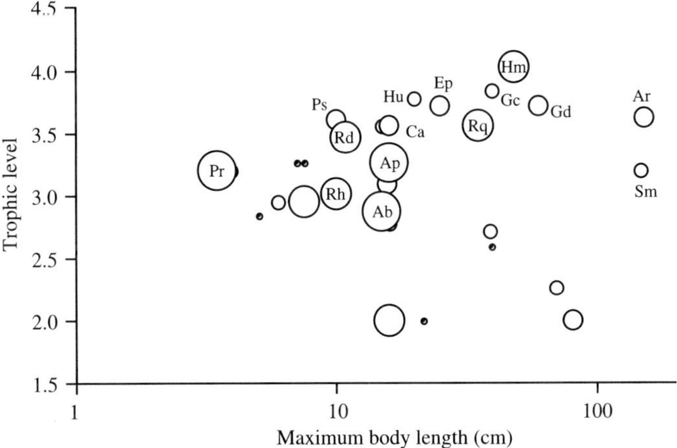

Fig. 2.4 Potential predators of guppies in Trinidad. The graph plots the trophic level of fish in the Trinidad assemblage against their maximum length (log 10 scale). These data are from Fishbase—www.fishbase.org: see also Table 2.1. The relative abundance of each species in Trinidad is denoted by the size of the symbol—the larger the circle the more widespread the species is. Five levels of incidence are shown: present in >50% sites in Trinidad; 25–49% sites; 5–24% sites, and <5% sites; not recorded in survey but known to exist in Trinidad. Incidence data are taken from the study described by Magurran and Phillip 2001. Key fish species are indicated as follows: Ab *A. bimaculatus*; Ap *A. pulcher*; Ar *A. rostrata*; Ca *C. alta*; Ep *E. pisonis*; Gc *G. carapo*; Gd *G. dormitor*; Hm *H. malabaricus*; Hu *H. unitaeniatus*; Pr *P. reticulata*; Ps *P. schomburgkii*; Rd *R. dayi*; Rh *R. hartii*; Rq *R. quelen*; Sm *S. marmoratus*. (See Table 2.1 for details of species)

are rare in Trinidad. Fig. 2.4 therefore also shows the relative incidence of fish species. Widespread species, feeding at higher trophic levels, and large enough to be effective predators of small fish, warrant attention as important candidate piscivores. *H. malabaricus* meets all the criteria for close consideration. Indeed, the impact of *H. malabaricus* on *Rivulus* populations is already extensively documented by Doug Fraser and Jim Gilliam who use the epithet 'strong piscivore' to describe its activities (Gilliam *et al.* 1993; Fraser *et al.* 1995, 1999; Gilliam and Fraser 2001). The consequences of *H. malabaricus* predation for guppy populations are less well understood, partly because the species is not abundant in the low order streams where the majority of guppy studies are focussed and partly because it is most active at night (Fraser *et al.* 1995). It clearly deserves more attention. Best known of the guppy predators, the pike cichlid *C. alta*, lies well within the piscivory domain of Fig. 2.4. Other species, not generally regarded as guppy predators, but worth investigating, include *Roeboides dayi, Polycentrus schomburgkii, Rhamdia quelen, Gymnotus carapo*, and *Anguilla rostrata*. Reznick and Bryga's (1996) decision to investigate predator-driven life-history evolution in guppy populations coexisting with *E. pisonis* and *G. dormitor* is well supported by this analysis.

There are four ways in which the predatory activity of fish is assessed. A classic approach is to examine a fish's stomach contents to determine what it has recently been eating. Alternatively, experiments can reveal predator preferences for particular sizes or types of prey and provide information on hunting tactics. Sometimes these experiments are supplemented by observations of fish in the wild. Third, behavioural assays can confirm that guppies react to potential predators, and determine how different predators are ranked in terms of perceived threat. Finally, mark-recapture and census techniques may be used to infer natural mortality rates.

Although stomach content analysis is a common technique in fisheries biology (Gerking 1994) few researchers have applied the approach to potential guppy predators. A rare exception is Ben Seghers's (1973) study of six fish species. Over 30 years on these data still provide our best insight into the feeding choices of Trinidadian fish. Fig. 2.5 supports the status of *C. alta* and *H. malabaricus* as important piscivores but does not prove—as Haskins *et al.* (1961, p. 380) asserted—that *Crenicichla* is quite possibly a 'specialized *Lebistes* (=guppy) predator'. In contrast, *Astyanax* and *Hemibrycon* are confirmed as omnivores though almost 20% of the *Astyanax* diet consists of guppies. *Rivulus* also consume guppies (around 10%) but eat predominately invertebrates, especially beetles (11%) and ants (42%). Like *Rivulus, Aequidens* is an invertebrate feeder and molluscs make up 42% of its diet; there is no indication from these results that it is piscivorous, though guppies may respond to it as a potential predator (see below). These data have been influential in guiding research on the guppy system over the past three decades. However, it is important to recognize, as Seghers himself (1973) does, that the sample sizes, particularly for the piscivores, which are numerically not abundant in the assemblage, are small. In a separate study Fraser and Gilliam (1995) examined the stomachs of 72 *Rivulus* and found that they contained mainly terrestrial and aquatic insects; none had consumed guppies.

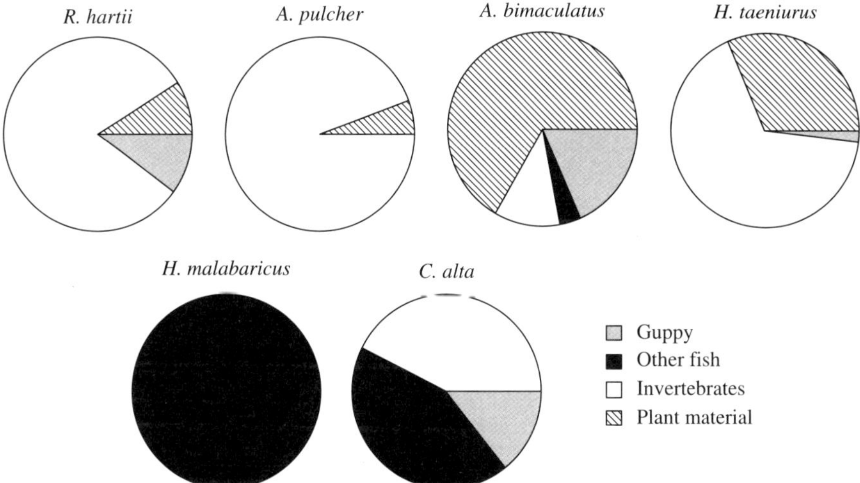

Fig. 2.5 Diets of some potential guppy predators. The pie diagrams show the percentage frequency of occurrence of four classes of food item: guppies; other fish; invertebrates; plant material (including algae). All fish were collected in Trinidad. Sample sizes were as follows: *R. hartii*, $n = 259$; *A. pulcher*, $n = 42$; *A. bimaculatus*, $n = 42$; *H. taeniurus*, $n = 64$; *H. malabaricus*, $n = 7$; *C. alta*, $n = 14$. Data are from Seghers (1973).

Investigations of the behaviour of predators, and their prey, are more plentiful. Haskins *et al.* (1961) placed guppies and a selection of predators in aquaria and monitored guppy survival. Their experiments confirmed that *Rivulus* and *Astyanax* attack and eat guppies. In both cases, female guppies survive longer. However, in neither case were the predators particularly effective. For example, when 100 female and 100 male guppies were placed together with four adult *Rivulus* in a 223 l aquarium, 50% of males and females were still alive after 14 days and 22 days, respectively. In contrast, *Crenicichla* proved to be efficient piscivores. Fifty per cent of individuals had been captured and eaten within 1–2 days. Haskins *et al.* (1961) concluded that colourful males are at a selective disadvantage relative to drab females. Indeed these data were used to underpin their conclusions regarding selection by predators on colour patterns in guppy populations in Trinidad. Seghers (1973) carried out similar experiments using only a single predator (as opposed to the four in the Haskins study) and taking greater care to size match the fish. He concluded that when body size is controlled, males are not necessarily more vulnerable to predators. And where the Haskins work is used to shore up conclusions regarding colour patterns and evolution the Seghers study has been used as evidence for size-selective predation on large prey. These two studies, with apparently contradictory conclusions inspired a number of follow-up investigations with equally inconsistent results. (In fact, when analysed in the same way the Haskins and Seghers studies

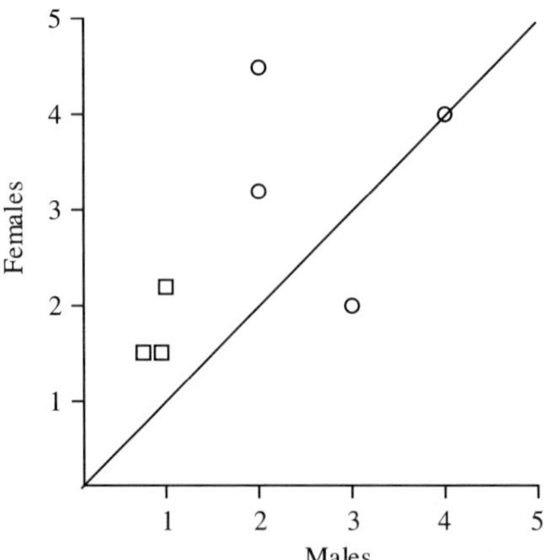

Fig. 2.6 Relative survival of male and female guppies exposed to *Crenicichla*. The graph summarizes results from two separate investigations. Haskins *et al.* (1961) added 50 male and 50 female guppies to a 223 l aquarium that also contained 4 *Crenicichla*. There were three replicates (square symbol). Seghers (1973) used the same number of guppies but had a 200 l aquarium and a single *Crenicichla*. In this case, there were four replicates (circle symbol). The time (in days) to 50% mortality of each sex is shown. The diagonal line assumes equal survival of males and females. In five out of the seven cases females survived longer.

jointly point towards a female advantage: Fig. 2.6). Mattingly and Butler (1994) found little support for size-selective predation whereas Pocklington and Dill (1995) discovered that *Crenicichla* selectively attacked (the larger) females in mixed-sex pairs. O'Steen *et al.* (2002) found no gender effect on survival. More recently Johansson *et al.* (2004) have shown that *Crenicichla* attack the largest guppies if given the simultaneous choice of either two or four prey sizes (Fig. 2.7). The details of these analyses, and the relative vulnerability of different sizes, sexes, or colours of fish, have an important bearing on investigations of life-history and colour pattern evolution (see Chapter 5). And while the manner in which *Crenicichla* selects guppies is still a matter of some debate the fact that it is an important predator of them is corroborated by these results.

Many guppy biologists have anecdotal accounts of predator–prey interactions in the wild but to my knowledge, there is only one study in which these have been quantified. Endler (1987) and a team of helpers observed five *Crenicichla* in a pool in the upper El Cedro River in Trinidad. They recorded an average of 2.5 attacks per hour in the middle of the day (1000–1400 h). This fell to 1.2 attacks per hour in the morning and evening (0800–1000 and 1400–1600 h, respectively). Endler (1987) further points out that visually conspicuous courtship elements, notably the sigmoid display, are reduced in favour of sneak mating attempts under high light levels, and argues that this represents an adaptive shift in behaviour during the time

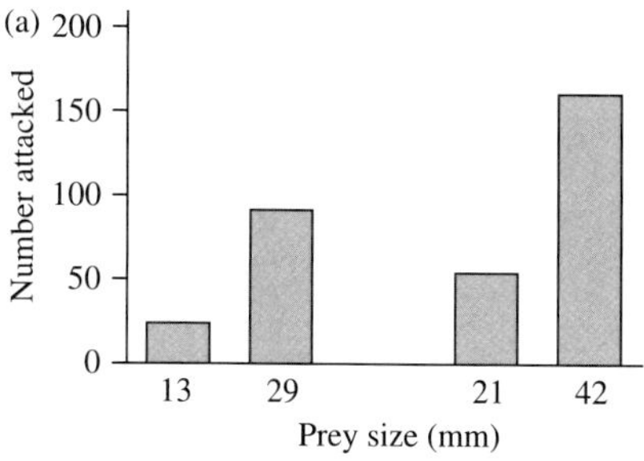

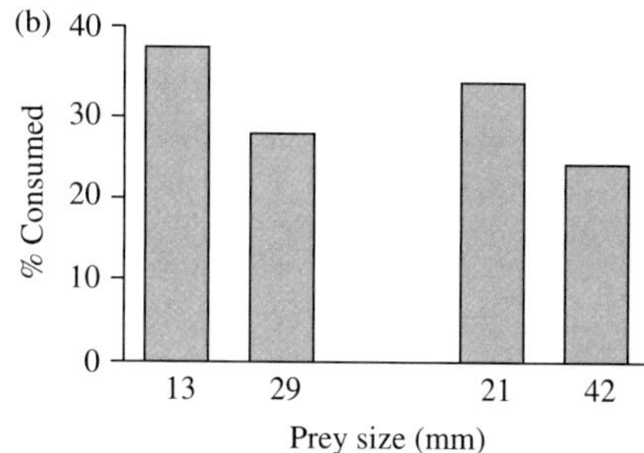

Fig. 2.7 Prey selections in *Crenicichla*. Pike cichlids were offered a choice between pairs of differently sized guppies—13 and 29 mm prey, and 21 and 42 mm prey. In both cases, the larger individuals were attacked more frequently (a). Although this translated into greater absolute numbers of the large prey being consumed, the percentage of attacked guppies that were eaten declined slightly for the larger member of the pair (b). Data are taken from fig. 6 in Johansson *et al*. 2004.

when predation risk is greatest. Other studies, for example, Reynolds *et al*. (1993), have confirmed that courtship behaviour in guppies is suppressed under bright light.

A number of investigators have quantified the behavioural responses that guppies make to potential predators. Fraser and Gilliam (1992) discovered that guppies avoided pools when *Hoplias* was present. Guppies that cannot move away from *Hoplias* have fewer opportunities to feed at night and devote less time to courtship during the day (Fraser *et al*. 2004). Seghers (1973) noted that guppies in a small lake were concentrated near the shore and attributed this to an anti-predator response towards large *Hoplias*. He contrasted this situation with the Paria River (which lacks piscivores) in which guppies are found in all parts of the stream apart from those areas where the water current is strong. Seghers (1974*b*) also reported a correlation between predation intensity and schooling tendency—a result that has been replicated in several other investigations (Farr 1975; Breden *et al*. 1987; Magurran and Seghers 1991). Repeated studies (e.g. Seghers 1973; O'Steen *et al*. 2002) have shown that guppies exhibit a vigorous escape response when they encounter *Crenicichla*. Other researchers have confirmed anti-predator behaviour in response to *Aequidens* (e.g. Magurran and Nowak 1991; Godin and Davis 1995*b*) and *Rivulus* (Seghers 1973; Magurran and Seghers 1990*a*). Relatively few investigators have attempted to discover how prey rank their predators. Seghers (1973) confirmed that there is a stronger response to *Crenicichla* than to *Rivulus* although Kelley and Magurran (2003*a*) did not find significant differences in the way in which guppies reacted to *Crenicichla* and *Aequidens* models, a result we attributed to subtle differences in the manufacture of the dummies. The perceived level of threat posed by various types of predator depends on the evolutionary history of the population in question, as well as the number of generations spent in captivity, and the experience of individual fish

(Seghers 1973; Magurran and Seghers 1990*a*; Magurran *et al.* 1992; Kelley and Magurran 2003*a*).

The final way of assessing predator activity is to monitor the fate of guppies experiencing different predation regimes. Reznick *et al.* (1996*b*) conducted mark recapture studies in three high-predation (primarily *Crenicichla*) and three low-predation (*Rivulus*) localities in Trinidad. Recapture probabilities were consistently lower in the *Crenicichla* sites. Direct estimates of adult mortality rates showed that there was 20% mortality per 12 days in high-predation sites as opposed to 10% mortality in low-predation ones.

2.2.2 Composition of fish assemblages

Although it is possible to find relatively simple fish communities in which guppies co-occur with a single predatory species, multiple predators are present in many assemblages (Magurran and Phillip 2001*a*). Risk is a function of community diversity. On average the probability that a serious fish predator, such as *Crenicichla* or *Hoplias*, will be present increases with assemblage size (Fig. 2.8). Thus guppies that occur in species-rich, which usually means downstream, localities are more likely to experience high-predation risk. However, it is not simply the presence of single predators that is important here but the interactions between different species in an assemblage. Predators may exclude one another and cause a net reduction in risk, or may alternatively divide the spatial and temporal guppy niche between them and thus limit the availability of predator-free refuges. Because of the convenient though simplistic low- versus high-predation classification of guppy populations, there is as yet only limited understanding of the consequences of assemblage structure for guppy evolutionary ecology.

2.2.3 Avian predators

Trinidad's rich avifauna makes it a favoured destination for birdwatchers. A total of 411 species have been recorded in Trinidad and at least 247 of these are known to have bred there (ffrench 1992). The list includes a number of species that are potentially important predators of guppies. For example, ffrench (1992) notes that five of the six American species of kingfisher, including the belted *Ceryle alcyon* and green *Chloroceryle americana* kingfishers, are present. In addition, there are 20 species of herons, egrets, and bitterns (Ardeidae and Cochleariidae) including the cattle egret *Bubulcus ibis*, which colonized Trinidad in 1951 and is now the most common heron on the island. Small fish form an important component of the diet of many of these species (ffrench 1992). Another near ubiquitous and emblematic bird is the kiskadee, *Pitangus sulphuratus*, a flycatcher. The kiskadee is a generalist feeder which consumes a wide variety of items including insects, small lizards, and birds, mice, fruit, and scraps (ffrench 1992). On three separate occasions, I have observed a kiskadee swoop down to a pool, capture and eat a guppy. Chadee *et al.* (1991) report that kiskadees can consume large quantities of guppies

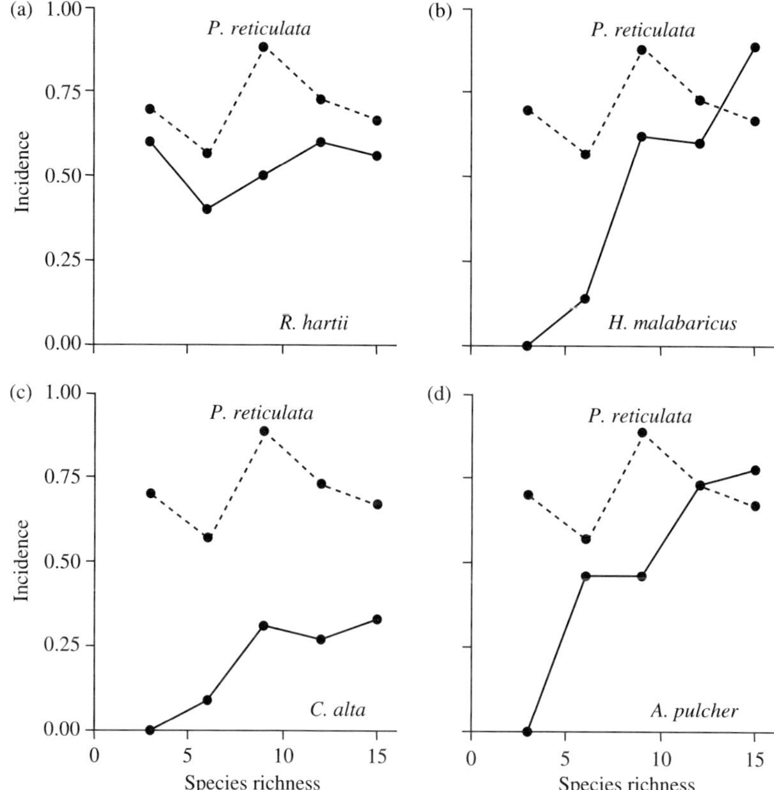

Fig. 2.8 Incidence function (Diamond 1975) of guppies (dotted line on all graphs) and (a) *R. hartii*, (b) *H. malabaricus*, (c) *C. alta*, and (d) *A. pulcher* in communities of different richness levels (1–3 species, 4–6 species, 7–9 species, 10–12 species, and 12–15 species). The incidence function plots the proportional incidence of species in assemblages of different sizes. After Figure 4 in Magurran and Phillip 2001.

and that feeding activity is greatest around mid-day (see Fig. 2.9). Kiskadees are abundant in rural as well as urban areas and can be found in plantations and forests up to an elevation of 450 m. Although they tend not to penetrate dense undergrowth or to occur under thick canopy their cosmopolitan habit and foraging behaviour makes them a potentially important guppy predator. Fish-eating anhingas (snake birds) and cormorants also occur in Trinidad but are rarely seen on the smaller rivers.

Despite this diversity, the role that birds play in the guppy predation story is rarely considered. Haskins (1961, p. 390) 'factored out birds as significant predators . . . because we only saw four herons in twelve years and no kingfishers'. Since then, few researchers have paid much attention to birds. However, some guppy populations do exhibit behaviour that is consistent with bird predation. For example,

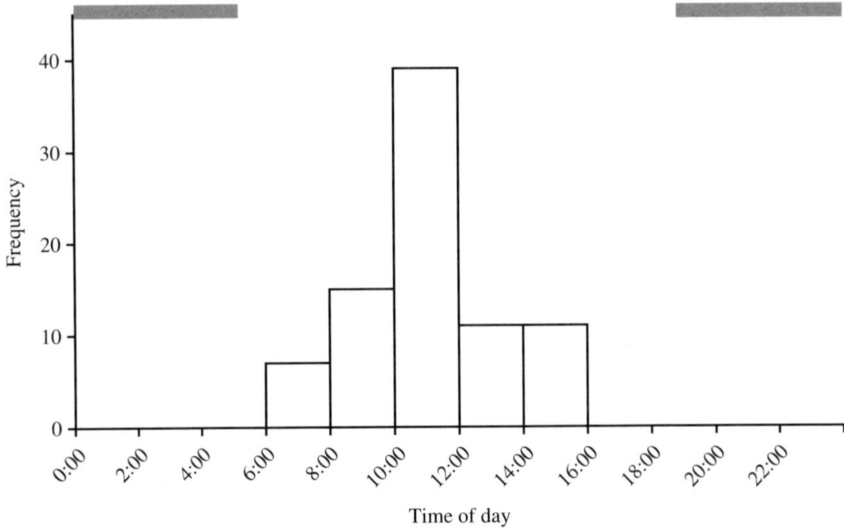

Fig. 2.9 Feeding activity of the Great Kiskadee (*P. sulphuratus*). The graph shows the number of visits to guppy holding tanks at St Joseph in Trinidad. During each visit the kiskadees were reported as consuming large quantities of fish. Feeding is most intense during the brightest hours of the day, and ceases during darkness (indicated by shading at top of graph). Data were collected in July 1988. The graph is based on table 1 in Chadee *et al.* (1991).

Seghers (1973, 1974*a*) found that guppies in the Paria river preferred deeper water and responded more vigorously to a model bird predator than fish in other populations.

One recent investigation has examined the dual impacts of aerial and aquatic predators (Templeton and Shriner 2004). Guppies collected from low- and high-(fish) predation sites (Marianne and Yarra, respectively) were exposed to a model of a green kingfisher (*C. americana*) or a live small-mouth bass (*Micropterus dolomieui*) a fish that resembles *Crenicichla*. Templeton and Shriner (2004) observed strong, but qualitatively different, responses to the two types of predators. There were also interesting population differences in reaction. Whereas Yarra guppies sheltered under cover in response to the model kingfisher, Marianne guppies usually froze on the bottom of the tank. It is interesting to note that Seghers (1973) observed a similar freeze reaction in guppies from the Paria river, a locality with an aquatic fauna that is similar to that of the upper sections of the Marianne. And, in confirmation of earlier work (e.g. Seghers 1973, 1974*a*; Magurran and Seghers 1990*a*), downstream Yarra guppies responded more vigorously to the fish predator than the upstream Marianne guppies did. Templeton and Shriner argue that the selection pressures from the two different types of predators have resulted in conflicting behavioural responses (see also Chapter 3).

2.2.4 Other vertebrates

Another aerial predator that may pose a significant threat to guppies is the fishing bat *Noctilio leporinus. N. leporinus*, which occurs in Trinidad (Seghers 1973) and in much of the region, has provided important insights into the echolocation of moving targets (Wenstrup and Suthers 1984; Hartley *et al.* 1989). The foraging behaviour of the species has also been investigated. Working on the island of Culebra, Puerto Rico, Brooke (1994) found that the fishing bat was both an insectivore and a piscivore and that pelagic and freshwater fish dominated the diet during the dry season. Schnitzler *et al.* (1994) discovered that fishing bats have several strategies for capturing fish. One approach is to search for prey within the vicinity of ripples in the water caused by jumping fish. Alternatively, a bat may drop to the water surface, lower its feet, and rake its claws through the water for a distance of about 10 m. Echolocation behaviour at this time suggests that the bats are engaging in a haphazard search for surface fish. Schnitzler *et al.* (1994) found that bats captured fish once in every 50–200 passes through the hunting area. Female fishing bats roost with the same individuals in the same location over long periods (Brooke 1997). These bats also repeatedly visit the same foraging area and communicate the location of prey to one another. Fishing bats may indeed be significant predators of guppies in Trinidad although their hunting behaviour suggests that they are unlikely to pose a major threat to populations found in the small fast-flowing streams of the Northern Range.

Other possible vertebrate piscivores include the spectacled caiman, *Caiman crocodilus*, and some of the many species of snake found in Trinidad. Although snakes have not been recorded as guppy predators in Trinidad, Kelley and Magurran (2003*a*) found that wild-caught guppies from the Lower Tacarigua were wary of a snake model. One possibility is that the fish treated the model as a swamp eel (*S. marmoratus*), which may be an occasional guppy predator (see above). Another is that snakes are indeed unrecognized predators of guppies. Snakes certainly predate small fish in other places. Macías-Garcia *et al.* (1994) examined the behaviour of the Mexican viviparous fish *Giardinichthys multiradiatus* in relation to its natural predator, the snake *Thamnophis melanogaster*. Guppies have been introduced into Mexico (Zambrano and Macías-Garcia 1999); they will encounter snake predators there as they must do through much of their current range.

2.2.5 Invertebrate predators

A striking feature of rivers in Trinidad, particularly those in northern (Antillean) drainages, is the abundance of freshwater prawns. Many of these prawns belong to the genus *Macrobrachium* (family Palaemonidae) which is common throughout the Caribbean (Chace and Hobbs 1969; Hunte 1978). *Macrobrachium* eggs hatch as free-swimming zoea and require saline water for successful larval development (Hunte 1978). This dependence on seawater implies a constant migration of juveniles and spawning females up- and downstream.

The species that are most abundant, and most likely to be guppy predators in Trinidad are *Macrobrachium carcinus*, *M. crenulatum*, *M. faustinum*, and *M. heterochirus* (I. W. Ramnarine, personal communication). However, the severity of this form of predation remains a matter of some debate. Endler (1983) reported that *M. crenulatum* appears to attack guppies at least as frequently as *R. hartii* does and categorized it as a moderately dangerous predator (Endler 1991). In contrast Luyten and Liley (1985) regarded *Macrobrachium* as at most 'a minor predator of the guppy'. Magurran and Seghers (1990*a*) showed that guppies from the Paria River, a site with high densities of *M. crenulatum*, avoided the 'attack cone'—risky zone around the mouth and chelipeds (see also Chapter 3)—of these prawns. Guppies from rivers where *Macrobrachium* are rare or absent were much less wary of this dangerous anterior region. Ben Seghers and I (unpublished study) also found that 23.6% of *M. crenulatum* prawns from the Paria and Yarra (Limon) rivers ($n = 168$) had guppy scales in their guts. However, in a small number of cases the estimated body length of the guppy (deduced from scale size) exceeded that of the prawn that had consumed it, suggesting that scavenging occurs, at least occasionally. Our investigation further uncovered a higher incidence of fin damage—which is incurred when prawns attempt to capture guppies—in sites were *Macrobrachium* are abundant. (Paria River 1989 ($n = 137$ adult guppies sampled); Paria River 1990 ($n = 210$); Yarra River 1990 ($n = 164$): $>15\%$ of individuals with fin damage in all cases. Turure River 1990 ($n = 200$): $<5\%$ with fin damage. Guppies in this section of the Turure River are exposed to a range of fish predators but encounter few prawns). These data suggest that *Macrobrachium* prawns can have at least a modest impact on guppy populations. However, Rodd and Reznick (1991) used discriminant function analysis to examine the life-history characteristics of female guppies found in *Macrobrachium* localities and concluded that either these prawns prey infrequently on guppies or do not select them on the basis of size. Mark-recapture data supported the conclusion that there was no size-selective predation.

The second area of interest concerns the selection that *Macrobrachium* prawns may exert on male coloration. It has been known for some time that male guppies in rivers with high densities of prawns—the Paria and Yarra Rivers being the most famous examples—have more orange and red markings on them than those found in other streams. This convergent coloration cannot readily be attributed to shared ancestry since Paria and Yarra guppies come from genetically divergent populations (Carvalho *et al.* 1991). Endler (1983, 1991) notes that *Macrobrachium* prawns are relatively insensitive to red colours and argues that caretonoid markings may provide a private wavelength in which males can safely signal to females. It is only by hunting during daylight that prawns can have any impact on male coloration. *M. crenulatum*, the species most likely to be active during the day, has received most attention from researchers (Endler 1983). *Macrobrachium* prawns are primarily crepuscular predators, however. After dark, densities increase dramatically as prawns emerge from their hiding places. A survey of the Paria River in January 1989, for example, found fewer than two *Macrobrachium* m^{-2} at the onset of dusk. One hour later, when it was completely dark, the density had risen to an average of 10 m^{-2} (B. H. Seghers and A. E. Magurran, unpublished data). Olfaction is likely to be more important than vision during night-time foraging. This raises a number of interesting possibilities. The low schooling tendency of Paria River guppies could, for example, be an adaptation that minimizes odour cues. Tactile cues may also be important. Size-selectivity might diminish under these circumstances. Much remains to be discovered about the foraging behaviour of these prawns and their impact on guppy behaviour and evolution.

Since large predatory fish often attack prawns as well as smaller fish (see, for example, Phillip 1993) guppies are unlikely to experience heavy predation from both types of predator at the same locality.

Other possible invertebrate predators include odonate larvae (families Zygoptera and Anisoptera) and the freshwater crab (*Pseudothelfusia garmani*) (D. A. T. Phillip, personal communication; Reznick *et al.* 2001*b*).

2.3 . . . and productivity

Water conditions just above and below a barrier waterfall—usually a distance of just a few metres—are often comparable. In practice, however, and for several important logistical reasons including access and the availability of habitats, such as pools, that support reasonable fish densities, guppy biologists often make contrasts between sites some distance above and below the barrier. For example, the Upper Aripo (Naranjo) and Lower Aripo site used in many investigations (including my own) are 5 km apart. Since the richness and abundance of predators often increase downstream, guppies in such localities are likely to experience greater risk than those immediately below a waterfall. However, predation risk is not the only factor that increases downstream and distance below a barrier is typically correlated with productivity. The relationship between productivity and phenotypic (as well as genotypic) variation in guppy traits is currently an active research area.

Reznick *et al.* (2001*b*) made a detailed study of high- and low-predation sites in Trinidad's Northern Range (see also Chapter 5). No significant differences in water quality or physical variables were detected, an unsurprising result given that the investigators deliberately chose pools that were similar in structure and differed only in predation status. High-predation localities tended to have more open canopies and higher light intensities. This translated into significant differences in gross periphyton production though not in net primary productivity. Macroinvertebrate density did not vary between predator communities. However, as Reznick *et al.* (2001*b*) point out, a random selection of high- and low-predation localities would be likely to differ much more markedly.

This qualification is supported by Magurran and Phillip's (2001*a*) survey of fish assemblages. We found guppies in every type of freshwater habitat in Trinidad, including rivers receiving domestic, agricultural, and industrial effluent. Variation in the amount of forest cover and temperature among sites (see Fig. 5.5) as well as in nutrient load, will translate into differences in productivity. Lowland rivers are also often naturally turbid and have different flow regimes from the mountainous streams typically investigated. Although little is known about the evolutionary ecology of guppies in these localities, the consequences of variation in productivity are well illustrated by an investigation of resource availability in six low-predation streams in the Northern Range (Grether *et al.* 2001*b*). Gregory Grether and his colleagues found that food availability for guppies increases as canopy cover decreases and more photosynthetically active light becomes available. They further demonstrated that cover explained 84% of variation among streams in algae availability, and that this in turn explained 93% of variation in guppy growth rates. Unicellular algae are the primary dietary source of the caretonoid pigments used to produce the red, orange, and yellow markings of males. These colour patterns are important in female choice (Kodric-Brown 1985, 1989; Houde 1987; Endler and Houde 1995) and are a correlate of condition (Endler 1980; Nicoletto 1991). Caretonoid availability is higher in rivers receiving more light (Grether *et al.* 1999, 2000) and, as such, likely to be a better indicator of male foraging ability in the shaded streams where there are fewer algae. However, Grether (2000) found only weak support for the hypothesis that the strength of female preference for caretonoid coloration is higher in localities where dietary caretonoids are limited. He suggests that female preferences may only loosely track the indicator value of male traits.

It is sometimes assumed that clear, oligotrophic mountain streams represent the primary habitat of the guppy and that its presence in eutrophic lowland waterways is an artefact of the disturbance caused by recent population growth and industrial development on the island of Trinidad. However, Regan (1906, p. 390) cites Mr Lechmere Guppy Jr. (son of the man who bequeathed his name to the species) as observing that 'This fish . . . is very plentiful, especially in such places as the "Dry River", at Belmont, a suburb of Port-of-Spain, where they swarm in the filthy soapy water that drains from the yards of the dwellings along the river'. To date few researchers have considered the effects of eutrophication and turbidity on guppy evolutionary ecology. The finding that there are quantitative shifts in the sexual behaviour of guppies inhabiting turbid lowland rivers (Luyten and Liley 1991) indicates that these effects could be substantial.

2.4 Feeding behaviour of the guppy

Guppies are primarily benthic feeders. Dussault and Kramer (1981) contrasted the feeding behaviour of male and female guppies from two localities—the Upper Aripo (Naranjo) and the Lower Tacarigua ('upstream' and 'downstream' sites, respectively). Algal remains, diatoms, and invertebrates dominated the diets. Guppies in the Upper Aripo consumed most invertebrates, which were mainly juvenile stages of aquatic insects. Lower Tacarigua guppies had more diatoms and mineral particles in their stomachs. With the exception of diatoms and algal remains in the Tacarigua, there were no differences between the sexes. (Fig. 2.10). This is unsurprising since male and female guppies share the same taste preferences (Nikolaeva and Kasumyan

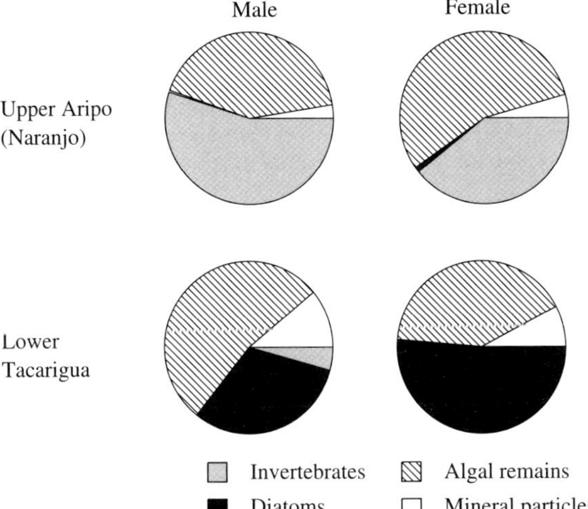

Fig. 2.10 Diets of wild guppies. These pie diagrams contrast the diet of male and female wild guppies from two localities—the upstream Upper Aripo (Naranjo) and the lowland Lower Tacarigua. Data are taken from table 2 in Dussault and Kramer (1981). Items that made up only a small fraction of the diets—higher plant fragments, filamentous and unicellular algae, and 'other' are not shown.

2000). As benthic algae are less nutritious than invertebrate prey (Dussault and Kramer 1981) it appears that guppies in the eutrophic Lower Tacarigua have poorer diets (though they have more access to caretonoids) than in the oligotrophic Upper Aripo. These population differences in diet probably reflect real differences in food availability at the two sites. Murdoch *et al.* (1975) for instance, demonstrated that guppies when offered two types of food would switch between them and consume whichever was most abundant. However, food availability is not simply the amount or type of food present in the habitat; it also encompasses the opportunities to consume particular food items and the risks entailed in doing so. One factor affecting food availability is the presence of competitors and a plausible explanation for Dussault and Kramer's result is that interspecific competition for invertebrate prey is higher in the species-rich Lower Tacarigua River. The extent, and consequences, of interspecific competition deserve much fuller investigation; most studies of feeding behaviour or foraging trade-offs in guppies simply ignore it. Predation risk also varies between the sites, and is known to have a strong influence on foraging. Godin and Smith (1988), for example, found that guppies that foraged in more rewarding food patches (in this case higher densities of zooplankton) were in greater danger of capture by a predatory cichlid. Milinski and Heller (1978) showed that three-spined sticklebacks (*Gasterosteus aculeatus*) preferred to feed in less profitable patches, and were more hesitant about attacking and consuming prey, in the presence of a predator.

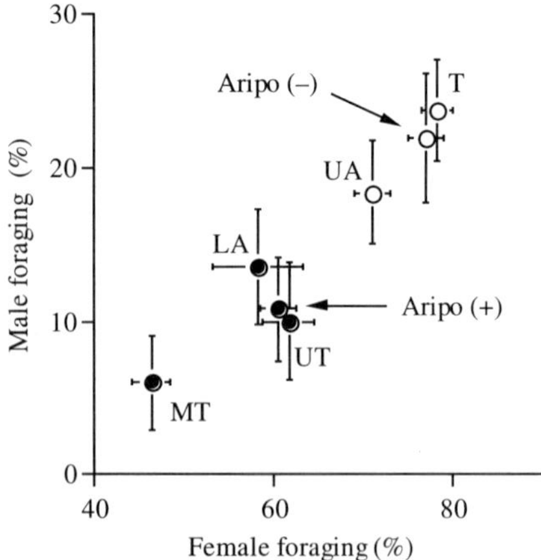

Fig. 2.11 Foraging effort by male and female guppies in the wild. This graph shows the mean percentage (±s.e.) of the time budget devoted to foraging by guppies in seven Trinidadian populations. High-predation ('*Crenicichla*') populations are represented by a filled circle, low-predation ('*Rivulus*') populations by an open circle. Aripo (−) is the tributary of the Aripo into which J. A. Endler transplanted guppies from the high-risk Lower Aripo (LA) in 1976. Aripo (+) is the previously predator-free section of the middle Aripo into which D. N. Reznick transplanted Crenicichla in 1981. The other populations are the Upper Aripo (Naranjo) (UA), Upper Tunapuna (T), Upper Tacarigua (UT), and Middle Tacarigua (MT). Data were collected in April 1992. The investigation is fully described in Magurran and Seghers (1994).

Although the diets of male and female guppies are similar, they differ in the trade-offs they make between foraging and other activities. Laboratory observations indicate that females feed at a rate about six times that of males, and that feeding rate is correlated with size in females but not in males (Dussault and Kramer 1981). The reasons for this are straightforward. Unlike males, females have indeterminate growth. Their fecundity is a function of body size (Hester 1964; Reznick 1983), which is itself a product of foraging success and age. These differences are reflected in the time budgets of wild fish (Fig. 2.11). Female guppies feed for much longer (by a factor of 2 or 3) than males in all habitats (Magurran and Seghers 1994*c*). Interestingly, these data also reveal that foraging occupies more of the time budget (of both sexes) in upstream, oligotrophic, and low-predation localities (see also Kolluru and Grether 2005). Dussault and Kramer (1981) further note that females were often courted while feeding whereas males do not court and feed simultaneously. Magurran and Seghers (1994*a*) discovered that wild female guppies subjected to persistent courtship by males reduce the amount of time they spend foraging (see also Chapters 4 and 6). Males and females reach different compromises between the risk of predation and choice of feeding patch (and potential energy intake) (Abrahams and Dill 1989) as well as between feeding and mating (Griffiths 1996).

A novel consequence of dietary preferences in guppies has recently been proposed by Rodd *et al.* (2002) who note that the bright orange fruits of the cabrehash tree,

Sloanea laurifolia, are rare but desired food items. Rodd *et al.* argue that the visual system of guppies is tuned to detect profitable orange food items, such as these fruits. Female preferences for large or intense orange colour spots in males could then have arisen as a result of the pre-existing sensory bias for orange objects. Rodd *et al.* find support for their hypothesis through the observation that both male and females are more responsive to orange objects than to objects of other colours. They also point out that variation in attraction to orange explains 94% of inter-population variation in female preference for orange. A further test would be to demonstrate that preference by females (and presumably males) for orange objects pre-dates the emergence of orange coloration in males (Basola 1990, 1995; Endler and Basolo 1998). Although this remains to be done it has been shown that females in some other poeciliid species are attracted to novel orange spots in conspecific males (Schlupp *et al.* 1999). Rodd *et al.* point out that *Poecilia picta* females do not appear to base mating decisions on orange coloration in males (Breden and Bertrand 1999). However, *P. picta* are usually found in rivers and pools near the coast—localities into which the cabrehash and other nutritious orange fruits are unlikely to fall. It would be interesting to examine the male and female preferences for orange in the guppies that also share these sites.

2.5 Parasites

Parasites are recognized as a major evolutionary driving force with their role in sexual selection attracting particular interest (Hamilton and Zuk 1982). It is, however, a topic that has received only modest attention where guppies are concerned. Wild populations are exposed to a diversity of parasites (Lyles 1990). Only one of these—the monogenean *Gyrodactylus*—has been studied in any detail. *Gyrodactylus* is an ectoparasitic worm with a direct life cycle. It is highly contagious and is transmitted directly between conspecifics (Scott and Anderson 1984; Scott 1985; Cable and Harris 2002; Cable *et al.* 2002). Two species have been recorded in Trinidad—*Gyrodactylus turnbulli* and *Gyrodactylus bullatarudis* (Lyles 1990; Richards and Chubb 1998; Oosterhout *et al.* 2003*a*). There are generally <10 worms per host fish but up to 50% of individuals in a population can be infected (Harris and Lyles 1992). Houde (1997) notes that guppies with fewer than 20 parasites generally show no sign of illness. Males infected with *Gyrodactylus* lose some of the brightness (chroma) in their orange spots (Houde and Torio 1992) and reduce the intensity of their courtship (Kennedy *et al.* 1987). Males that show greater immunity to infection have higher rates of sexual display, and are preferred by females. (López 1998). However, if females themselves are infected their preferences weaken and they become less discriminating (López 1999). Recently, Oosterhout *et al.* (2003*a*) uncovered population differences in the immune response to *Gyrodactylus*. One interpretation of their observation that upstream (Upper Aripo) guppies experience a higher and more prolonged parasite burden than downstream (Lower Aripo) is related to the low genetic diversity of isolated headwater populations (Oosterhout *et al.* 2003*a* and see also Fig. 2.3). It is possible that the major histocompatibility complex (MHC) (Sato *et al.* 1996) is implicated in this. The consequences of infestations by other parasites, and of interspecific competition among parasite species, are not well understood. McMinn (1990) documented adverse effects of nematode parasites in a feral population of guppies in Oxford University's Botanic Garden. Ornamental guppies are known to be subject to a range of parasitic diseases (e.g. Hatai *et al.* 2001; Thilakarante *et al.* 2003) that could also infect wild stocks or be transmitted to other species during introductions (see Chapter 7).

2.6 Dynamical aspects of ecology—variation over time as well as space

Population and community ecologists have long sought to understand and model stability and change (e.g. Nicholson and Bailey 1935; Elton 1958; Williams 1964*a*; MacArthur and Wilson 1967; May 1974; Lande *et al.* 2003; Southwood *et al.* 2003). However, the focus on spatial variation in predation regimes and productivity means that relatively little attention has been paid to temporal aspects of guppy evolutionary ecology. I use the examples of density and sex ratio to illustrate that it warrants consideration alongside geographical variation.

2.6.1 Density

Guppy densities vary considerably over space and time. Reznick and Endler (1982) found that guppies tended to have lower densities in high-predation localities (median value of two guppies m^{-2} for *Crenicichla* localities and nine guppies m^{-2} in *Rivulus* localities). In contrast, there was no significant difference between the numbers of guppies in high- and low-predation pools in Reznick and coworkers' (2001*b*) study (see Fig. 2.12). Reznick *et al.* (2001*b*) attribute this difference to the fact that some of high-predation localities in the earlier study were large streams. When data are presented in terms of biomass the difference between predation regimes is dramatic (Reznick *et al.* 2001*b*). Low-predation sites have four times more guppy biomass per unit volume than high-predation pools (mean values are 530 mg m^{-3} and 126 mg m^{-3}, respectively, dry weight measures). This difference is explained by the fact that there are proportionally more large fish in low-predation localities. Although productivity is associated with growth rate, Grether (2001*b*) found no consistent relationship between guppy biomass and canopy cover in six low-predation streams.

There may be fewer grams of fish per m^3 in high-predation sites but what about the guppy's eye-view of density? Guppies in high-predation sites have a high schooling tendency (Seghers 1974*b*; Magurran and Seghers 1991) and form shoals that typically consist of around 4–40 individuals (Croft *et al.* 2003*b*; Russell *et al.* 2004). Croft *et al.* (2003*b*) reported a mean density of 12 (s.d. = 7.7) guppies m^{-2} in a section of the Arima River in which shoals were present. This investigation also revealed that encounters between a focal individual and other individuals or shoals occur on average every 14 s (s.d. = 11 s).

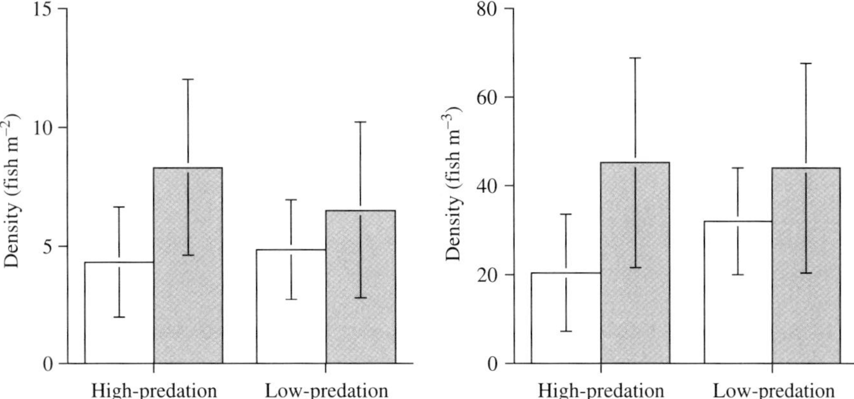

Fig. 2.12 Mean density (±1 s.e.) of guppies in high- and low-predation pools in Trinidad's Northern Range. The top graph shows density per unit area, the lower one density per unit volume. Open bars represent guppies >12 mm in length, stippled bars show guppies of all sizes. Data from table 4 in Reznick *et al.* (2001).

These summary statistics mask considerable variation. My own experience of visiting the same rivers in Trinidad in successive years is of being unable to predict, from one field trip to the next, how abundant the fish will be. In some cases, I arrive to find that a river has been perturbed or polluted (see also Chapter 7). In others, the canopy will have closed over, or alternatively trees may have been felled. There are further density changes that are not readily explained by habitat alteration. Fig. 2.13 illustrates this by showing the number of adult guppies, per m^2, in 20 small pools of the Upper Tunapuna River. These pools were censused completely (with replacement) during the dry season of four successive years. Even though the pools in this study remained broadly the same size and structure over the duration of the study, and were in undisturbed secondary forest, the fish densities fluctuated markedly. Flooding is one factor that can lead to a sudden change in density. Grether *et al.* (2001*b*) found that guppy biomass in low-predation streams fell between 22 and 92% after flooding. Drought, changes in the biotic community and in primary productivity, siltation, and disturbance are others. There are also marked seasonal shifts in guppy abundance (Reznick 1989). Life-history consequences of variation in density are reviewed in Chapter 5.

2.6.2 *Sex ratio*

Sex ratios in Trinidadian guppy populations are often female biased, particularly in low-predation localities (Haskins *et al.* 1961; Seghers 1973; Liley and Seghers 1975). Liley and Seghers (1975) proposed that males are vulnerable to predation by *Rivulus*—whereas the indeterminate growth of females protects larger individuals from being eaten. Subsequent work supports this pattern (e.g. Rodd and Reznick 1997), though has not clarified the mechanism, which is still in doubt. Male guppies devote more time

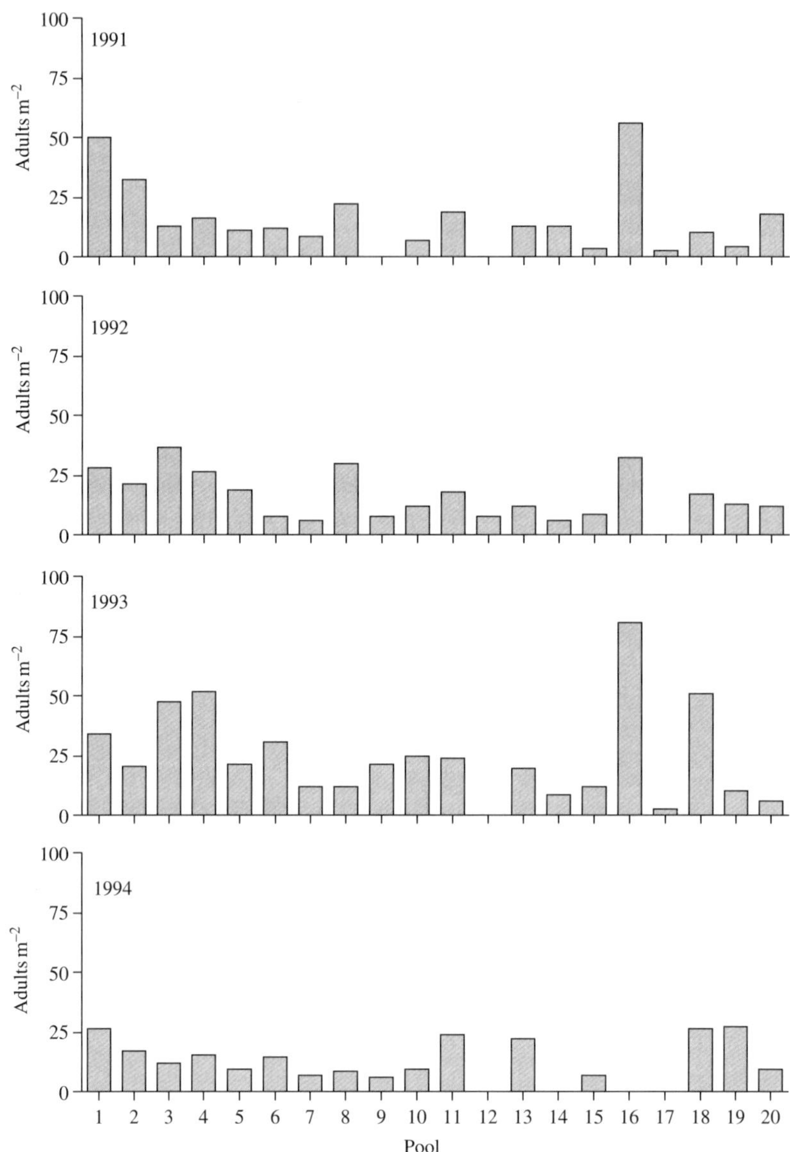

Fig. 2.13 Variation in guppy density over time in the Tunapuna River. The same 20 small pools, which form a linear series separated by riffles, were censused completely during the dry season (late March/early April) in four successive years. Pool 1 is furthest upstream; the length of the section of river censused is 82 m. Data presented are numbers of adults per m^2. Pool area ranged from 0.3 to 7.1 m^{-2}. Median size was 1.9 m^{-2}. Total number of adult fish per pool ranged from 0 to 359. Repeated measures ANOVA shows that there are significant differences among years in the densities of fish per pool ($F_{3,57} = 4.63$, $P = 0.006$). Data collected by B. H. Seghers and A. E. Magurran as part of an unpublished study.

to courtship and less to predator defence (Magurran and Nowak 1991; Magurran and Seghers 1994*c*)—and have poorer evasion skills than females (Seghers 1973 and see Table 3.3) so it seems plausible that predation is important. Male poeciliids also appear to be more prone to physiological stresses (Snelson 1989), including temperature extremes, food shortages, and overcrowding (Krumholz 1948). Immunocompetence may be a further contributory factor (Folstad and Karter 1992), particularly for the bright males in low-predation sites (which are already probably disadvantaged by reduced heterozygosity (Oosterhout *et al.* 2003*a*)).

Changes in sex ratio have repercussions for reproductive behaviour and sexual conflict. Farr (1976, 1980*b*) showed that male sexual activity is mediated by the number of competing males in the vicinity. Jirotkul (1999*a*) documented significant shifts in male mating tactics as the sex ratio varied from 0.17 to 0.83 (where this is the proportion of males). Males devote more time to courtship and increase their sigmoid display rate in female biased populations. In contrast, sneak copulations become more prominent, and male–male competition increases (Jirotkul 1999*b*) if there is a male surplus. Females are choosier under a male-biased sex ratio as there are more potential mates (Jirotkul 1999*a*) and the opportunity for sexual selection increases (Jirotkul 2000*b*). Jirotkul's experiments show vividly how changes in operational sex ratio mediate reproductive interactions. But as Jirotkul used receptive females in her work, the outcome does not necessarily reflect the situation in the wild. Here only a small fraction of females will be sexually receptive at any given time (Magurran and Seghers 1994*c*) so even female-dominated populations could be effectively male biased in terms of the true operational sex ratio. An added complication is that males adjust their response to the receptivity status of the female as well as the intensity of male–male competition. Early experience further modifies behaviour. The sex ratio that males experience as they develop influences their sexual activity as adults. Evans and Magurran (1999*b*) reared males in male-biased, female-biased, and evenly balanced groups. Fish in the male-biased rearing treatment performed relatively more sneaky mating attempts. This effect persisted even when they were tested in a 1:1 sex ratio.

There is considerable potential, then, for behaviour to be moulded by shifts in local sex ratio, and for the arena in which sexual selection operates to change. These shifts can be substantial. Pettersson *et al.* (2004) recorded sex ratios at two upstream (low-predation) and two downstream (high-predation) localities over a 12-month period (Fig. 2.14). There were considerable fluctuations with both female bias and male bias being recorded on occasions. Rather than a convergence on a stable sex ratio all populations displayed long-term oscillations that are consistent with recent models of sex ratio dynamics (Caswell and Weeks 1986; Lindström and Kokko 1998; Ranta *et al.* 2000; Pen and Weissing 2002). One of the populations in Fig. 2.14—Tunapuna—had previously been exhaustively censused on an annual basis. These data (Fig. 2.15) reinforce the conclusion that sex ratios vary over time.

It is now well known that many species have the capacity to influence the sex of their offspring. Offspring sex allocation is essentially a dynamic process (Caswell and Weeks 1986; Byholm *et al.* 2002) since parents accrue fitness advantages from

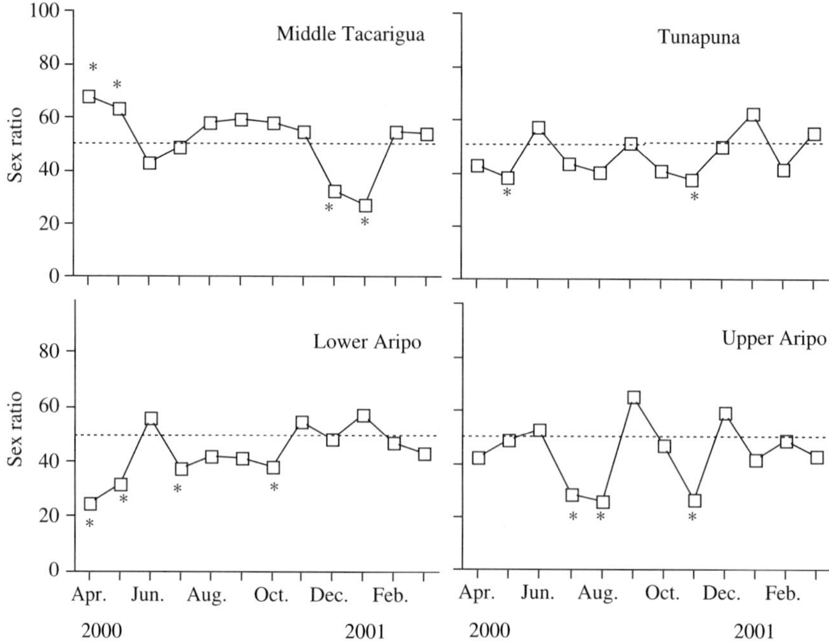

Fig. 2.14 Adult sex ratios over 12 successive months, beginning April 2000, in two sections of the Aripo River in Trinidad, and in two sections of the Tacarigua system. The Upper Aripo and Tunapuna sites are 'low-predation' localities, their partner sites are 'high-predation' localities. Adult fish (at least 100) were collected using a one-person seine and dip nets, sexed and released. Data are presented as percentage males. The dotted line marks the equal sex ratio. Significant deviations from this are denoted by asterisks. All four populations show significant variation over time. See Pettersson *et al.* (2004) for further details.

biasing their offspring towards the rarer sex (Fisher 1958; Trivers and Willard 1973). Fisherian (1958) models of sex ratio dynamics predict that juvenile sex ratios will compensate for deficits in the adult population. Pettersson *et al.* (2004) examined adult and juvenile (see Takahashi 1975) sex ratios in 11 Trinidadian populations. These populations are geographically dispersed—they include the Pilote River in the SE corner of the island, as well as localities in the Northern Range—and have varying water quality and predation regime. There is no evidence of a negative correlation between adult and juvenile sex ratios and thus no support for a Fisherian compensating process. Indeed, this survey was consistent with previous investigations (Haskins *et al.* 1961; Seghers 1973) which concluded that guppy sex ratios at birth do not deviate from 1:1. It also supports the rebuttal (Brown 1982) of earlier work (Geodakyan *et al.* 1967; Geodakyan and Kosobutskii 1969) that claimed that guppies could regulate the sex of their broods in a compensatory fashion. (Biased sex ratios in inbred strains of guppies are a different matter—see Chapter 7 for a discussion).

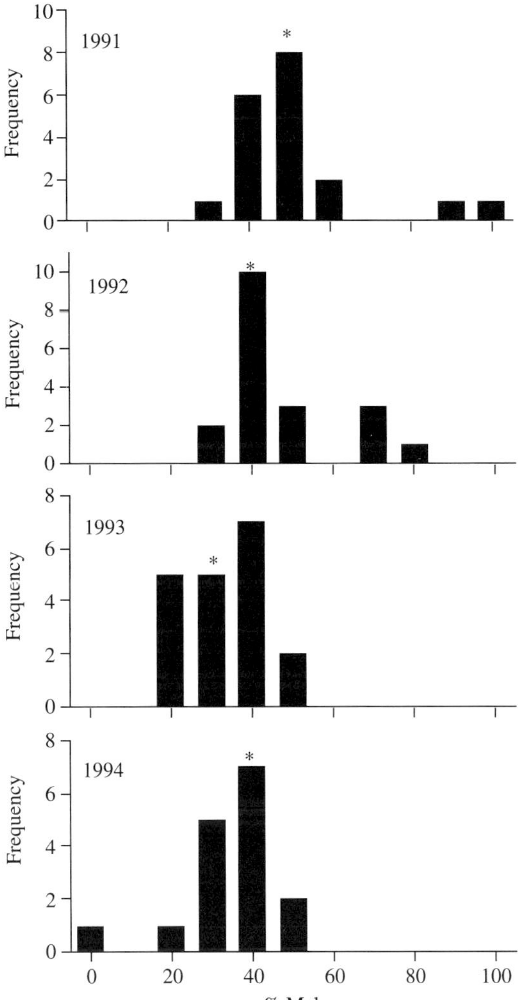

Fig. 2.15 Frequency distribution of sex ratios in the 20 pools of the Upper Tunapuna over four successive dry seasons (1991–1994). * indicates the class in which the median pool lies. The system becomes more female biased over time. (Repeated measures ANOVA $F_{3,45} = 6.74$, $P = 0.0008$). See Fig. 2.13 for details of the site and sampling programme.

2.7 Geographical variation in guppy traits

Endler (1995) tabulated almost 50 guppy traits that covary with predation intensity. It is this rich variation that has fuelled research and attracts increasing numbers of researchers to Trinidad. It also forms the subject matter of much of the remainder of

Table 2.2 Summary of differences between guppy populations in low- and high-predation localities

Traits	Characteristics of low-predation populations relative to high-predation ones	Reference
Morphology (males)	Larger, more brightly coloured	(Haskins *et al.* 1961; Endler 1983); Chapters 3 and 4.
Anti-predator responses	Weaker	(Seghers 1974*b*) Table 3.1
Sexual selection	More intense	(Houde and Endler 1990) Chapter 4
Sexual conflict	Possibly weaker	(Magurran 2001) Chapter 6
Life history	Later maturity, reduced reproductive allocation	(Reznick and Endler 1982) Chapter 5
Life expectancy in the wild (females)	Greater	(Reznick *et al.* 1996*b*)
Life expectancy in the laboratory (females)	No differences	(Reznick *et al.* 2004)
Parasite diversity	Lower	(Lyles 1990; Harris and Lyles 1992)
Parasite resistance	Lower	(Oosterhout *et al.* 2003*a*)
Molecular diversity	Lower	(Carvalho *et al.* 1991)
Level of interspecific competition	Possibly lower	(Dussault and Kramer 1981)

the book. Table 2.2 summarizes the differences between guppy populations in low- and high-predation localities, mindful of the fact that this geographic variation may be influenced by environmental factors that correlate with predation, or shaped by the indirect effects of predation.

2.8 Conclusions

The ecology of the guppy in Trinidad is generally interpreted in terms of the contrast between low- and high-predation localities, where high-predation sites are defined as those in which the pike cichlid *C. alta* is present. However, the situation is not as clear cut as generally assumed and complexities in predator–guppy interactions are widely ignored in the literature. There are four matters arising here. First, other predators, in addition to pike cichlids, and encompassing birds, mammals, reptiles, and invertebrates, as well as fish, are undoubtedly implicated in guppy evolutionary ecology. But, with a few notable exceptions, their impact is poorly understood. Second, we know virtually nothing about the relationship between the diversity of an assemblage and the manner in which natural selection is expressed. I mean this both in terms of the interactions between putative predators and in the context of interspecific competition. There will also be indirect effects of predation and competition to consider. Third, because low-predation sites are almost always in headwaters and high-predation ones in the downstream sections of rivers, predation risk is typically confounded by other variables, such as productivity, temperature, and water flow. Investigators have sought to control these confounding effects in various ways, in order to demonstrate that predators alone can act as agents of selection. The interactions between predation risk and other environmental variables, and the manner in which these jointly shape evolution is nonetheless an important, if neglected, topic. Finally, the fact that guppies breed continuously throughout the year may be responsible for the emphasis on spatial rather than temporal variation in density, sex ratio, population structure, predator–prey interactions, and behaviour. The few studies that have addressed the dynamical features of guppy ecology indicate that this is a missed opportunity.

3

Evading predators

Predator–prey relationships, notably those involving groups of animals, have been authoritatively reviewed by a recent predecessor in this series (*Living in groups* by Krause and Ruxton 2002). I recommend their book to the reader searching for a comprehensive and insightful review of the field. But whereas the strength of Krause and Ruxton's monograph is its breadth of coverage and integration of studies dealing with a range of functions of grouping behaviour across diverse taxa, investigations of the guppy illuminate the diverse consequences of predator avoidance and living in groups within a single species. It is notable that many of the examples cited in *Living in groups* are drawn from the guppy literature. Guppy research offers a near unique combination of systematic field observations, controlled laboratory experiments, and insights into the heritability and evolution of anti-predator responses. Indeed Trinidadian guppies provided one of the first experimental demonstrations that predators have a significant impact on behaviour and morphology. The significance of this research extends beyond its contribution to our fundamental understanding of predator–prey interactions, however. Overall, the single greatest bequest of the guppy to evolutionary ecology is to reveal the direct and indirect consequences of modifications to one set of traits. In order to survive, guppies must make behavioural and morphological adjustments to predation risk. These are linked to life-history changes and this suite of anti-predator adaptations in turn has implications for sexual selection and reproductive behaviour.

Although Caryl Haskins (Haskins *et al.* 1961) drew attention to the pivotal role of predators in the guppy evolution story he had relatively little to say about behavioural responses. Seghers (1973, 1974*b*) was the first person to describe guppy anti-predator behaviour in detail and to quantify population differences in response. Since then, there has been a flurry of investigations. For example, one recent focus of research, with some surprising outcomes, has been the physiological mechanisms that underpin anti-predator responses (Odell 2002; Odell *et al.* 2003). This chapter will begin with a brief general introduction to predator–prey interactions. The consequences of variation in predation risk for Trinidadian guppies, and the trade-offs linked to effective predator defences, will then be evaluated. When can adaptive differences be classed as evolutionary change? What are the pitfalls associated with such assumptions? Populations vary in their predisposition to acquire anti-predator behaviours. Does this represent a difference in learning skills? Age-related changes in morphology and behaviour will be explored. Finally, the chapter will examine differences between the sexes in response to predation. This will set the scene for the next chapter.

3.1 Predator–prey interactions

The nature of predator–prey interactions in fish has been comprehensively discussed elsewhere (see, for example, Godin 1986; Pitcher and Parrish 1993; Fuiman and Magurran 1994). In essence, effective prey responses depend on appropriate behaviour at all six stages of the predator–prey sequence. An effective approach is simply to *avoid* risk by occupying predator-free space. However, assuming that this is not possible, the prey need to *detect* a potential risk, such as a different odour or the presence of a larger fish or predatory bird in the vicinity. The capture rate for prey that have not detected a predator is much greater than for those that have already responded (Webb and Skadsen 1980; Webb 1982). Next, it is crucial to *discriminate* between a genuine threat and a benign encounter. Is the intruder a piscivore and, if so, is it hungry or actively hunting? Failure to respond to a real predator may result in death but equally too many false alarms will detract from courtship and foraging opportunities. At this point, the prey has the option to adopt behaviours that will *inhibit* the predator from attacking while continuing to *monitor* its behaviour. Finally, should the predator attack, the prey may engage in various *evasion* manoeuvres. Learning allows anti-predator skills to be honed throughout an individual's lifetime. Heritable individual variation in anti-predator responses or the predisposition to learn about predators is the raw material on which natural selection operates. However, the simplest method of surviving predation is to *avoid* encounters with a predator in the first place. This section therefore begins by discussing the behavioural and morphological adaptations that guppies employ to distance themselves from predatory attack.

3.1.1 Predator avoidance

Changes in microhabitat use are a simple but effective means of reducing risk of predation. Seghers (1973) noted that guppies in high-predation localities were concentrated in shallower water at the edges of the rivers, whereas those found in less dangerous habitats also exploited deeper water and the middle sections of streams. He further mentions that guppies that moved away from their shore refuge in the Lower Aripo River were immediately pursued by predators. These observations were supported by controlled tests on laboratory stocks of guppies (Seghers 1973). Odell (2002) made focal observations of wild fish and found that guppies in low-predation (upstream) localities occupied significantly deeper water (29 ± 2 cm versus 21 ± 2 cm (mean $\pm$ s.e.), used sites with faster flowing water (14.2 ± 1.7 versus 4.8 ± 0.7 cm s^{-1}), and spent a larger percentage of their time being active (58 ± 5 versus 34 ± 4) than fish in downstream sites. A footnote to these studies is my personal observation that larger females in high-predation localities are sometimes found offshore, in faster flowing water. This raises the interesting possibility that females may trade-off risk of predation against persistent courtship by males, which by virtue of their smaller size are less able to maintain station in faster currents (see also Croft *et al.* 2004*b*).

Diel shifts in behaviour are another method of reducing risk of predation. Fraser *et al.* (2004) found that guppies continued to forage at night in the absence of nocturnal predators, particularly *Hoplias*. As *Crenicichla* primarily hunt during daylight hours (but see Seghers 1973 for evidence that they can successfully hunt after dark) one option for guppies exposed to diurnal but not nocturnal predators would be to redirect some of their foraging activity to nights. Interestingly, suppression of nighttime foraging leads to reduced courtship activity during the day (see Chapter 4) and has implications for growth and the evolution of life histories (see Chapter 5) (Fraser *et al.* 2004). A further temporal modification of behaviour was recorded by Endler (1987) who noted that guppy courtship is suppressed under high light levels, a shift associated with an increase in *Crenicichla* hunting activity around midday (see also Reynolds *et al.* 1993). Despite these intriguing results, few investigators have attempted to follow behaviour or examine predator–prey interactions throughout the diurnal cycle. Doug Fraser and his colleagues (Fraser *et al.* 2004) make an important point when they say that the temporal activity patterns of guppies (and other species) deserve much more attention.

The cryptic beige-grey coloration of female guppies appears to offer protection against predators. Although there has been a number of studies comparing the vulnerability of males and females (see Chapter 2) to my knowledge no one has confirmed that female morphology is protective when other variables, such as behaviour and size are factored out. Male colour patterns vary systematically across populations with fish in more dangerous localities being generally less brightly patterned (Haskins *et al.* 1961). Endler (1980) confirmed that when males are released from predation colour spots become larger and more numerous and colour patterns more diverse. He further noted a link between gravel size and spot size; background matching occurs in sites where predation is severe whereas the colour patterns of males that are the target of female choice but not of predators are much more conspicuous. Only a fraction of guppy populations in Trinidad are found in streams with clear water and uniform gravel bottoms. In mountainous areas, the stream bed may be composed of rock or sand, or covered with vegetation, while lowland rivers can be naturally turbid and frequently have a muddy substratum. My students have sometimes remarked on the striking coloration of male guppies collected in lowland, high-predation sites. It is possible that these bright colour patterns represent a different compromise in the trade-off between natural selection (predation risk) and colour pattern (sexual selection). Males in lowland areas are also usually much smaller than those found in the Northern Range (Alkins-Koo 2000 and see Chapter 5). Alternatively, sympatry with the congeneric *Poecilia picta* may select for particular colour combinations (see also Chapter 6). Investigations of lowland populations have the potential to extend our understanding of the link between female choice, predation risk, and male coloration.

3.1.2 Detection

There is ample evidence to indicate that guppies, like other fish, respond adaptively to potentially threatening stimuli. For example Fraser and Gilliam (1987) and

Abrahams and Dill (1989) found that aspects of the foraging behaviour of guppies changed in the presence of predators. Other investigations (e.g. Magurran and Nowak 1991; Godin and Briggs 1996; Gong 1997; Evans *et al.* 2002*a*) have confirmed that both male and female reproductive behaviour is altered when a predator is in the vicinity.

As is evident from the role that male colour patterns play in female choice, guppies have excellent colour vision (Endler 1991). Anstis *et al.* (1998) report that wild fish are 50% more sensitive to short wavelengths and 67% more sensitive to medium wavelengths than human observers. They also respond to ultraviolet wavelengths (Kodric-Brown and Johnson 2002; Smith *et al.* 2002). Less is known about the visual acuity of guppies. Experiments on inspection behaviour (Dugatkin and Alfieri 1992; Magurran *et al.* 1992) confirm that guppies can detect predators that are 0.5 m or more away while Seghers (1973) found that the maximum reaction distance (see below) to a predator model was 90 cm. A comprehensive analysis of visual capabilities is, however, still awaited. Vogel and Beauchamp (1999) point out that the reaction distance in fish is a function of light intensity and turbidity and these variables, as well as the size, coloration, and behaviour of a predator will determine whether a guppy can detect it in enough time to respond. Seghers (1973) also uncovered geographic variation in reaction distance. His study measured the distance at which guppies in five laboratory stocks of guppies (descended from Trinidadian populations) responded to a preserved 190 mm *Crenicichla*. Lower Aripo (high-predation) guppies reacted to a moving predator at a mean distance of 33 cm whereas those from the (low-predation) Petite Curucaye responded when it was only 17 cm away. Overall, there was broad correspondence between the level of predation risk experienced by the ancestors of these fish in the wild and their reaction distances, though some fish, for example, those from the (low-predation) Upper Aripo (Naranjo), were more wary than expected.

It is now clear that guppies are alerted to the presence of potential predators by olfactory cues (Nordell 1998; Brown and Godin 1999) as well as by visual ones (Kelley and Magurran 2003*a*). The lateral line system enables fish to discriminate objects that move at different speeds, or differ in size or shape (Vogel and Bleckmann 2000). Its role in predator detection or discrimination in guppies, or indeed other fish has, however, received little attention.

3.1.3 Discrimination

As Chapter 2 revealed, there are still relatively few studies demonstrating that guppies can rank potential predators in terms of risk. Magurran and Seghers (1990*a*) found that wild-caught guppies showed greatest attack-cone avoidance (avoidance of dangerous mouth region) of the type of predator that posed the highest risk to their population. For example, guppies from the Paria River—where decapod crustaceans are thought to act as predators—were more cautious in the presence of a freshwater prawn (*Macrobrachium crenulatum*) than guppies from the Lower Aripo River. Kelley and Magurran (2003*a*) found that wild-caught individuals kept a greater distance from models of two cichlid predators (*Crenicichla alta* and *Aequidens pulcher*) than a generic model of a snake (snakes are not thought to be important predators of guppies in Trinidad, see Section 2.2.4). This study also confirmed that fish from high-predation localities respond more strongly to predator models than fish from low-predation environments. However, these differences were muted when guppies were raised through two generations in the laboratory. This result suggests that early experience differentially mediates the anti-predator responses of fish that originate from a high-predation locality.

An intriguing study by Licht (1989) showed that guppies displayed a stronger anti-predator response to a hungry predator as opposed to a satiated one and moreover that the discrimination was greater in fish derived from a high-predation locality (Lower versus Upper Turure). Licht used a non-native predator—the largemouth bass *Micropterus salmoides*—in his investigations. Given the small number of studies on predator discrimination it would be revealing to extend this work.

3.1.4 Inhibition

Fish that have not detected a predator are extremely vulnerable to capture. For example, Krause and Godin (1996) found that a blue acara cichlid (*Aequidens pulcher*) preferred to attack guppies that had shown no response to them. The fact that predators are less successful when they attack wary individuals means that prey fish have an opportunity to signal their vigilance to predators. However, the extent to which prey engage in such pursuit deterrence behaviour remains controversial. I (Magurran 1990*a*) showed that pike (*Esox lucius*) were less likely to attack European minnows (*Phoxinus phoxinus*) that engaged in inspection behaviour. (The primary function of inspection behaviour appears to be predator monitoring—see below) Godin and Davis (1995*b*) extended this approach to demonstrate that the risk of attack (by *A. pulcher*) on inspecting guppies is lower than on non-inspecting ones. Experiments of this type are open to criticism as being correlational since predators could be monitoring the condition of the fish rather than their behaviour *per se* (Godin and Davis 1995*a*; Milinski and Boltshauser 1995). In practice it is extremely difficult to design experiments that effectively tease apart signals that advertise perception from those that reveal the individual's condition (Caro 1995). Moreover, it appears that predator inhibition is not the primary function of inspection behaviour (Magurran 1990*a*). However, the possibility that predators glean information from approaching fish about their prey's preparedness to flee as well as its ability to flee, remains tantalizing. I hope that someone will be stimulated to devise the definitive experiment on this putative function of inspection behaviour.

Fish may also on occasion mob potential predators (Dominey 1983; Dugatkin and Godin 1992*b*). The first report I am aware of is by Day (1880, p. 47) who comments that 'as small birds mob those of prey, so little fish will mob others that they dread. Some small species were kept by Mr Whitmee in an aquarium with an *Antennarius* (a frogfish) and were evidently in dread of their carnivorous neighbour, which they continually tried to torment. In attacking it they always took care to strike at its posterior part, although this was protected by a rock of coral'. This sounds similar to the attack-cone avoidance behaviour described earlier for guppies. Despite the early interest in mobbing behaviour by fish, few investigators (Hein 1996 is a rare example) have attempted to study it. To the best of my knowledge, it has not been reported for guppies.

Grouping is a behaviour that confers a variety of anti-predator benefits on its participants (Krause and Ruxton 2002). A definitive study by Neill and Cullen (1974) showed that the success rate per attack by predators was reduced when they attacked schools of prey rather than solitary individuals. One reason for this is the confusion effect, whereby the predator finds it difficult to single out an individual prey when faced with multiple choices. The classic analogy is with a child who cannot choose among the options presented by a box of chocolates but who will rapidly consume a sweet if only one is offered. The perceptual difficulties of targeting a single prey are enhanced when individuals cluster in a compact group (Milinski 1990)—a response often seen in fish schools (see, for example, Magurran and Pitcher 1987). Milinski and Heller (1978) demonstrated that predators that are themselves at risk of predation attack smaller aggregations of prey. Because of the confusion effect, and the related anti-predator benefits of schooling (see Box 3.1), it is possible that schooling has an inhibitory effect on predators. Krause and Godin (1995) tested this idea using guppy shoals of various sizes. When a predator (*A. pulcher*) was presented with a binary choice of shoal sizes (the guppies were shielded by a one way mirror and could not see the predator) it consistently 'attacked' the larger group. Manipulations of guppy activity (achieved by varying water temperature) showed that it was conspicuousness rather than shoal size *per se* that guided the predator's behaviour. And when the predators were offered free-ranging shoals they tended to attack the closest group of fish. These observations offer little support for the thesis that schools are relatively protected against attack though they do not negate the dilution effect which offers a per capita benefit to fish in schools (Pitcher and Parrish 1993; Krause and Ruxton 2002).

3.1.5 Predator monitoring

Although the pursuit deterrence function of inspection remains unproven there is now compelling evidence that prey use this behaviour to gather information about their predators. Magurran and Girling (1986) found that minnows (*P. phoxinus*) actively inspected pike (*E. lucius*) models that varied in shape and marking. Unrealistic models initially received more attention but were ultimately treated less cautiously than realistic ones. Interestingly, inspecting fish swam along the sides of the predator models (Fig. 3.1). Webb (1982) points out that pike are rounded in cross section and that the dorsal fin is located towards the posterior end of the fish. This shape makes it difficult for prey to identify approaching predators. However, the absence of an anterior median fin means that the body form of pike and their esocid relatives is suboptimal for acceleration lunges. Detectability is apparently traded off against strike efficiency. This trade-off could explain why prey fish often examine

Box 3.1 Shoals and schools

The primary function of schooling, a behaviour that is widespread in fish (Breder 1951; Keenleyside 1955; Radakov 1973; Shaw 1978; Pavlov and Kasumyan 2000), is predator evasion (Godin 1986; Pitcher 1986; Pitcher and Parrish 1993). It has long been known that schooling tendency is stronger in fish species that are more at risk of predation. Pelagic marine species, for example, form the large cohesive schools that provide memorable instances of synchronized and coordinated evasion tactics (Shaw 1962). The link between level of risk and degree of schooling was confirmed when Seghers (1974*b*) demonstrated that Trinidadian guppies from populations subject to intense predation had a well-developed schooling response. As these guppies had been raised in the laboratory and were predator-naïve, this work was also important in showing that there can be heritable variation in behaviour among populations of a single species. Subsequent laboratory (Breden *et al.* 1987) and field studies (Magurran and Seghers 1991) provided further support for Seghers's conclusions.

Pitcher (1983) makes a useful distinction between the terms 'shoal' and 'school'. Shoals are defined as social (rather than sexual) groupings of fish, analogous to 'flocks' of birds. This is distinct from the 'aggregation' that is formed when individuals are attracted to a defined area or a common resource but not to one another (Williams 1964*b*). 'Schools' are a type of shoal in which individuals show coordinated swimming behaviour or engage in synchronized manoeuvres. Following Pitcher, most behavioural researchers now use the general term shoal to refer to the social units with which they work. However, the terms 'schooling tendency' and 'shoaling tendency' meaning (usually) the time that a focal individual or individuals spends associating with conspecifics are often used interchangeably. 'Schooling intensity', a related measure, is the comparison between the size of group formed by a number of individuals and the size of group that would result if those individuals moved at random (Williams 1964*b*). It is similar to the notion of an 'elective group size' which measures the distribution of shoal sizes in free-ranging fish (see, for example, Magurran and Pitcher 1987), though usually without comparing this against a random expectation.

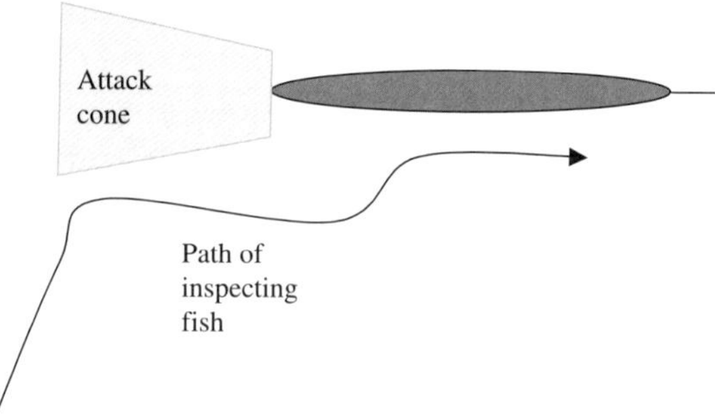

Fig. 3.1 Path taken by an inspecting fish towards a potential predator. The attack cone around the mouth of the predator is avoided, but the inspector typically swims along its flank, probably to confirm its identity using visual and olfactory cues. Tactile (lateral line) information may also be available.

the lateral profile of a potential predator. Another explanation for approach to the flank of the predator is that prey fish may use their lateral lines to obtain information, for example, on opercular rate, or predator shape or movement (Vogel and Bleckmann 2000), that might enable them to assess readiness to strike. Pike cichlids share the elongate shape of their Holarctic namesake. It would thus be intriguing to ask whether they have made parallel morphological trade-offs and to investigate the manner in which guppies obtain and evaluate information about this important predator. Paxton *et al.* (1994) proposed that a function of caudal ocelli (eyespots on the tail of a fish) in pike cichlids is to confuse prey (including guppies) and thereby gain an advantage during attack. An experimental test of this idea showed that more inspections were directed towards the tail of a model pike cichlid that sported an ocellus.

3.1.6 Predator inspection and reciprocity—a guppy's eye view

As noted above, the primary function of inspection appears to be predator monitoring. Predator inspection has been adopted as a model system for studying the evolution of cooperation (Milinski 1987; Dugatkin 1988; Milinski *et al.* 1990*a*, *b*; Dugatkin 1991*a*, *b*; Dugatkin and Alfieri 1991*a*, *b*). It remains open to question whether the apparent cooperation of inspecting fish provides evidence for reciprocity; research on guppies has fuelled both sides of the argument. Dugatkin (1997) provides a fascinating account of the topic, and reviews the relevant guppy literature. In brief, it is assumed that inspectors may find themselves in a prisoner's dilemma. This means that, during an inspection, pairs of fish have various options and that there are costs associated with each of these options. They may, for example, decide to inspect together. In this case, the heightened cost of proximity to the predator, **R**, is shared by both partners. If fish A, however, decides to wait behind while its partner inspects, it potentially benefits from information about the predator without the need to risk attack. The cost of doing this is **T**. On the other hand, fish A may choose to inspect alone. By doing so it must bear cost **S**. Finally, if neither fish inspects, they share the loss of information entailed, which is assumed to cost **P**. A prisoner's dilemma is assumed if the following inequality holds: $T > R > P > S$ (ranked so that S has the highest cost). Much of the discussion has been directed towards assessing the veracity of these assumptions. Because inspection is dangerous (Dugatkin 1992*b*; Milinski *et al.* 1997) it is difficult to reject the assumption that $T > S$. Similarly Milinski *et al.* (1997) confirmed that the risk of two inspectors is diluted so that $R > S$. Demonstrating that $T > R$ and $P > S$ is the key to showing that inspecting fish are indeed in a prisoner's dilemma rather than in a situation where benefits accrue as a consequence of 'no-cost cooperation', a state of affairs usually termed 'by-product mutualism' (Dugatkin 1997). By-product mutualism would, for example, arise if individual fish obtain more benefits from inspecting than by observing inspectors at a distance. We know that fish watching an inspector can acquire information about a predator, and that they modify their behaviour as a result, albeit not exactly in the manner that they would have done had they inspected themselves (Magurran and

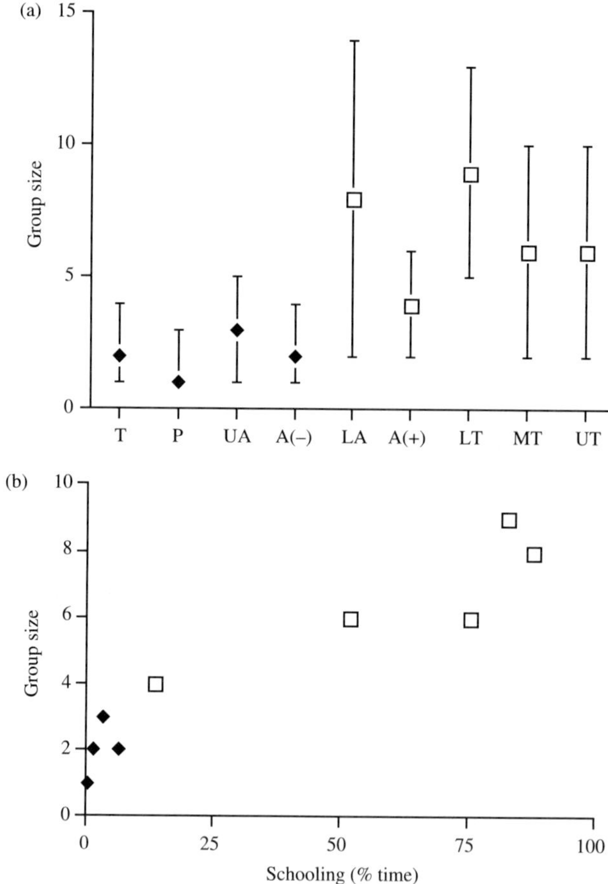

Fig. 3.2 Inspection behaviour in the wild. (a) This graph shows the median group size (plus interquartile range) in which the typical guppy inspects a model *Crenicichla*. Sites where *Crenicichla* presently occur are shown as an open square; those from which they are absent (and where *Rivulus* are found) are denoted as a filled diamond. The probability of an individual guppy inspecting a predator as a member of a group of two is greatest in the populations experiencing the least risk. In contrast, guppies that co-occur with *Crenicichla* are more likely to inspect in larger groups. The populations examined were as follows: T Upper Tunapuna; P Paria; UA Upper Aripo (Naranjo); A(−) Aripo tributary (i.e. low-predation site into which LA guppies were transplanted: see Table 3.2); LA Lower Aripo; A(+) Middle Aripo (i.e. site where *Crenicichla* was introduced: see Table 3.2); LT Lower Tacarigua; MT Middle Tacarigua; UT Upper Tacarigua. The grid references of these sites (with the exception of the Paria site, which was located downstream of Brasso Seco village) are provided by Magurran and Seghers (1994*c*). Data are taken from Magurran and Seghers (1994*b*) which also describes the methods in detail. (b) This graph plots median group size during inspection (as in the above) against mean schooling tendency for the population (based on time budget data gathered for 30 females per population—see Magurran and Seghers (1994*c*) for details of method). The two measures are strongly correlated ($r_s = 0.95$, $P < 0.01$). In other words, guppies from populations that school more inspect potential predators in larger groups.

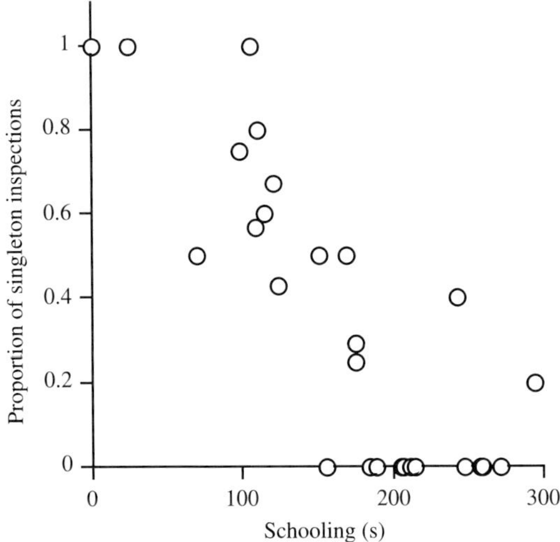

Fig. 3.3 Female guppies with a higher schooling tendency are less likely to engage in singleton inspections. This graph shows the proportion of singleton inspections made by an individual guppy, housed in a tank with four companion females, towards a 14 cm model *Crenicichla* (the same type of model was used in the study presented in Figs. 3.2 and 3.4). This is plotted against the time the same individual spent schooling (out of a maximum of 300 s) during a separate test when no predator was presented. There is a strong inverse relationship between these variables ($r = -0.83$, $P < 0.001$).

Higham 1988). It seems likely that non-inspectors gain imperfect information—perhaps because they are limited to visual cues and miss out on tactile and olfactory ones. What is uncertain is whether this imperfect information is good enough to make P > S. Designing experiments that conclusively resolve these issues is a formidable challenge.

Predator inspection may be undertaken by a single fish but singleton inspections are relatively infrequent, particularly in populations that school more readily (Figs. 3.2 and 3.3). Interestingly, inspections by pairs of fish, the group size most conducive to the emergence of co-operative behaviour, are also uncommon, most noticeably so in fish in localities where there are predators—that is precisely those places in which co-operative inspection might be expected to be favoured (Fig. 3.2(a)). The baseline level of schooling in a population appears to be a good predictor of group formation during inspection. This implies that, for the most part, inspection behaviour is underpinned by by-product mutualism rather than by reciprocal altruism. In other words, school members are gaining a safety in numbers benefit rather than cooperating *per se*. However, this does not preclude reciprocity between pairs of fish. Female guppies seem the most likely candidates since they display stronger site fidelity (Croft *et al.* 2003*a*), a greater tendency to associate with familiar individuals (Griffiths and Magurran 1998) and engage in more anti-predator behaviour (Magurran and Nowak 1991) than males.

3.1.7 Evasion tactics

Seghers (1973) described the evasion tactics of guppies under predator attack. In addition to 'avoidance drift' that is broadly similar to the inspection behaviour discussed above, he recorded 'turn around' (an abrupt reversal of swimming direction), 'rapid dart' (burst swimming), as well as 'surface skim' and 'surface jump' (where the guppy swims just below the water surface or leaves the water one or more times in quick succession). He also noted that these responses were elicited more readily in populations co-occurring with predators. My own observations confirm that guppies engage in the co-ordinated escape responses, such as the flash expansion and fountain manouevres (Potts 1970; Nursall 1973; Pitcher and Wyche 1983; Magurran and Pitcher 1987) typically displayed by schooling fish. These may even be seen in newborn fish (Magurran and Seghers 1990*b*). Time–budget analyses of wild fish indicate that evasion tactics are displayed more frequently by individuals (especially females) in high-predation localities (Magurran and Seghers 1994*c*).

Certain aspects of guppy escape behaviour have come under detailed scrutiny. Seghers (1973) showed that wild-caught guppies originating from upstream (low-predation) localities were more vulnerable to attack by *Crenicichla* and *Hoplias* than their downstream counterparts. These differences persisted in predator-naïve stocks that have been raised under controlled laboratory conditions for *c*.5 generations. O'Steen *et al.* (2002) also found that wild-caught guppies from high-predation populations were better at surviving staged encounters with predators. Common garden experiments again confirmed that these population differences in escape ability have a genetic basis. Odell *et al.* (2003) examined the physiology underlying aerobic capacity and burst speed and found, contrary to expectation, that there were no clear differences between predation regimes. It appears that the marked contrast among populations in life history and behaviour is not associated with changes in enzyme function or organ size. Indeed, Odell and his colleagues noted that observed differences in performance were explained by body size and concluded their paper with the comment 'guppies from high-predation downstream habitats are probably, on average, slower in absolute burst speed than their larger counterparts from upstream low-predation sites'. These results hint that learning and experience are important in determining survival. The contribution that learning makes to anti-predator behaviour is reviewed later in the chapter.

3.2 Consequences of variation in predation risk

As the preceding sections have made clear, there is marked variation in guppy anti-predator responses, much of it related to the geography of risk. Fish from dangerous localities are better at detecting, discriminating, assessing, and avoiding predators than those from sites where predation intensity is relaxed. Table 3.1 catalogues guppy anti-predator traits, summarizes differences that correlate with predation regime and indicates studies where a genetic basis for this variation has been uncovered. Although

Table 3.1 Phenotypic differences among populations

Trait	High-risk '*Crenicichla*' and/or '*Hoplias*' populations	Low-risk '*Rivulus*' populations	
Microhabitat use	Shallower, slower water	Deeper, faster water	(Seghers 1973; Odell 2002)
Diet activity	Cease feeding at night in the presence of nocturnal predators	Continue foraging at night	(Fraser *et al.* 2004)
Coloration	Males more colourful		(Haskins *et al.* 1961; Endler 1980)
Courtship behaviour	More sneaking	More display	(Endler 1987; Magurran and Seghers 1990*c*; Godin 1995)
			(Farr 1975; Rodd and Sokolowski 1995)
Discrimination of hungry versus satiated predator	Stronger		(Licht 1989)
Attack-cone avoidance	Preferentially avoid mouth region of predator	Weaker avoidance of predator mouth, closer approach	(Magurran and Seghers 1990*a*)
Group size during inspection	Larger		(Magurran and Seghers 1994*b*)
Cooperation	More cooperative		(Dugatkin and Alfieri 1992)
Approach distance during inspection	More wary		(Magurran and Seghers 1994*b*)
Schooling tendency	Higher		(Seghers 1974*b*; Breden *et al.* 1987; Magurran and Seghers 1994*b*)
Reaction distance	Greater reaction distance		(Seghers 1973)
Escape responses	Stronger		(Seghers 1973)
Evasion tactics	Quantitative differences in deployment of evasion tactics		(Seghers 1973)
Escape ability	More likely to survive attack		(Seghers 1973; O'Steen *et al.* 2002)
Burst swimming	No difference		(Odell *et al.* 2003) (Chappell and Odell 2004)
Aerobic performance	No difference		(Odell *et al.* 2003) (Chappell and Odell 2004)

Notes: This table lists some of the traits directly implicated in predator avoidance and indicates how they vary between predation regimes. For details of modifications of courtship behaviour and life history traits see Chapters 4 and 5.

guppies are no different from other species of fish in the types of anti-predator tactics deployed, they provide a particularly nice example of the way in which suites of tactics track variation in risk. Moreover, the system illustrates the broader consequences of that variation. Predation risk affects much more than immediate anti-predator responses, and as later sections of the book will reveal, there are few aspects of guppy evolutionary ecology that are not directly or indirectly influenced by it. For example, males modify their courtship tactics, not simply as a result of their own responses to predators (Endler 1987; Magurran and Seghers 1990*c*), but also in reaction to the females' heightened preoccupation with predator avoidance (Magurran and Nowak 1991; Evans *et al.* 2002*a*). This shift in courtship behaviour and the network of responses that accompany it, are investigated in Chapter 4. Life-history changes, such as the timing of maturity, and the pattern of senescence are explored in Chapter 5. In addition, heritable variation in behaviour and other anti-predator traits is the raw material of evolution. As the next section demonstrates, this evolution can occur over remarkably short timescales.

3.3 Evidence for evolution

Caryl Haskins was ahead of his time in realizing that a simple way to demonstrate that predators are responsible for evolution is to transplant fish from one predation regime to another and observe their fate. This approach was a natural extension of his research on gene flow in which he used natural colour markers to track the progress of males and their progeny up and downstream (Haskins *et al.* 1961). In 1957, Haskins moved guppies from a high-predation locality in the Caroni drainage to a low-predation one, the Upper Turure, in the Oropouche drainage. This excerpt from a letter sent to B. H. Seghers on 29 September 1992, following our rediscovery (Magurran *et al.* 1992; Shaw *et al.* 1992) of the transplant, provides some interesting background.

> The original aim of that introduction was to ascertain what happened to the male population percentage of fish carrying a conspicuous iridescent blue color marker, ('SB') when transferred from a population with high-predation, (the lower Arima), to a virtually predator-free environment. Finding such an environment at that time, of course, required extensive search, and when we found one that seemed to fill the bill, we examined it quite thoroughly for predators at intervals over a four-year period (1953–1957) before undertaking the introduction. Except for the occasional (and expected) *Rivulus hartii* we found no potential predators in the Turure from considerably above the 'rock wall' all the way down to the bridge spanning the stream just above the series of barrier waterfalls separating the upper from the lower stream, except for a single rather large specimen of *Hoplias malabaricus* one year, which seemed to have surmounted those barrier waterfalls well downstream. During all those searches, we never saw a single guppy in the river.

Haskins further notes that when the introduction (of 200 adult founders) was made, the population increased rapidly and soon 'saturated' the main river and two of its minor tributaries. Although Haskins records the source of the founders as the Arima

Table 3.2 Documented introduction experiments in Trinidad

Site of introduction, grid reference, and drainage	Source population, grid reference, and drainage	Type of manipulation	Date	
Upper Turure PS 997 817 Oropouche drainage	Lower Guanapo PS 913 765 Caroni drainage	*c*.200 guppies moved from high-predation 'Crenicichla' to low-predation 'Rivulus' locality	1957	(Shaw *et al.* 1992; Russell *et al.*, in review)
Aripo tributary PS 931 800 Caroni drainage	Lower Aripo PS 942 778 Caroni drainage	*c*.200 guppies moved from high-predation 'Crenicichla' to low-predation 'Rivulus' locality	1976	(Endler 1980)
El Cedro upstream PS 893 794 Caroni drainage	El Cedro downstream PS 893 794 Caroni drainage	*c*.100 guppies moved from high-predation 'Crenicichla' to low-predation 'Rivulus' locality	1981	(Reznick and Bryga 1987)
Aripo midstream PS 936 798 Caroni drainage	—	Crenicichla introduced into low-predation guppy population	1981	(O'Steen *et al.* 2002.

River, at the Churchill-Roosevelt Highway Bridge, subsequent research, using microsatellites, suggests that the source was actually the Guanapo River—which also crosses the highway a short distance away (see Russell 2004—this result is consistent with Shaw *et al.* 1992). In the event Haskins monitored frequencies of the *Sb* gene in the Turure for several years but found the results difficult to interpret and did not publish them. (The details of this and other guppy transplants are provided in Table 3.2).

John Endler carried out the next, and much better-known, introduction of guppies in Trinidad almost two decades later (Endler 1980). Like Haskins, Endler located a stream, separated from the main river by a series of waterfalls, where only one species of fish, *R. hartii*, was found. (This site is often referred to as Aripo I in the literature). Endler introduced a sample of 'about 200 guppies', collected from the Lower Aripo (high predation), into this tributary in July 1976. He monitored them in the following December, by which time the fish had spread up- and downstream, and again in early May 1978. The result was intriguing. In less than 2 years the colour patterns of the descendants of introduced fish had come to resemble males from the low-risk Naranjo (Upper Aripo) River. The sizes of the black, carotenoid, and iridescent spots had increased, colour diversity had increased and the area of the body covered by colour spots was greater. Endler suggested that this time period represented 15 generations of relaxed predation pressure. Later estimates imply that generations are more protracted (Reznick *et al.* (1997) calculate a mean 1.74 generations per year for fish in low-predation localities) making the outcome even more remarkable. This investigation proved that male colour patterns (and the genes they express) are shaped by natural selection, in the form of predation risk, and by sexual selection, in the form of female choice. Later experiments have helped unravel the way in which sexual selection operates against a background of variable predation risk (see Chapter 4 for details).

Endler's study highlighted the power of the transplant experiment to demonstrate rapid evolution in the wild. Subsequent research by David Reznick (reviewed in Chapter 5) revealed that predators were responsible for predictable shifts in life-history traits. The first demonstration that a change in predation regime resulted in heritable differences in anti-predator behaviour came when my colleagues and I made the serendipitous discovery that guppies in the Turure River were descended from another population. This came about when we were engaged in a population genetic study of Trinidadian rivers. Initially, using allozymes, we observed marked divergence between the Caroni and Oropouche drainages (Carvalho *et al.* 1991)—a result consistent with a mtDNA-based analysis (Fajan and Breden 1992). However, when we extended our survey we made the perplexing observation that the Turure River, though geographically part of the Oropouche drainage belonged, genetically speaking, to the Caroni drainage (Shaw *et al.* 1991). Correspondence with Caryl Haskins (personal communication 1990, 1992, and see above) soon resolved the problem and provided an opportunity to investigate the genetic consequences of an artificial introduction (Shaw *et al.* 1992 and see Chapter 6), and to determine how anti-predator behaviour had been affected by a relaxed predation regime.

In order to minimize the effects of environmental variation my colleagues and I (Magurran *et al.* 1992) examined the behaviour of fish that had been raised under standardized laboratory conditions for at least two generations. Fortuitously, because the Lower Arima was severely disturbed at the time of the collections, we used fish from the Lower Guanapo to represent the source population. Our observations revealed a highly significant drop in schooling tendency in the descendants of the introduced population (Fig. 3.4(a)). We also found that the inspection behaviour of guppies in the Upper Turure was in line with that typically observed in a low-predation stream (Fig. 3.4(b)).

O'Steen *et al.* (2002) took this approach one stage further and examined the escape behaviour of guppies in a series of matched pairs of populations. In each case guppies from high-predation and low-predation localities within a stream were compared. Three of the comparisons involved the descendants of an artificial introduction and their natural source. Two natural population pairs were included as controls. These fish were tested as wild-caught individuals. For three of population pairs guppies were also raised under common garden conditions in the laboratory, and tested at the F_2 generation. During a trial 12 size-matched guppies, 6 from each population in a comparison were placed in a small pool that contained a *Crenicichla*. Red and black marks were used to denote the origin of the guppies. Trials continued until about six guppies had been captured (a period of between 15 and 240 min); the identity of

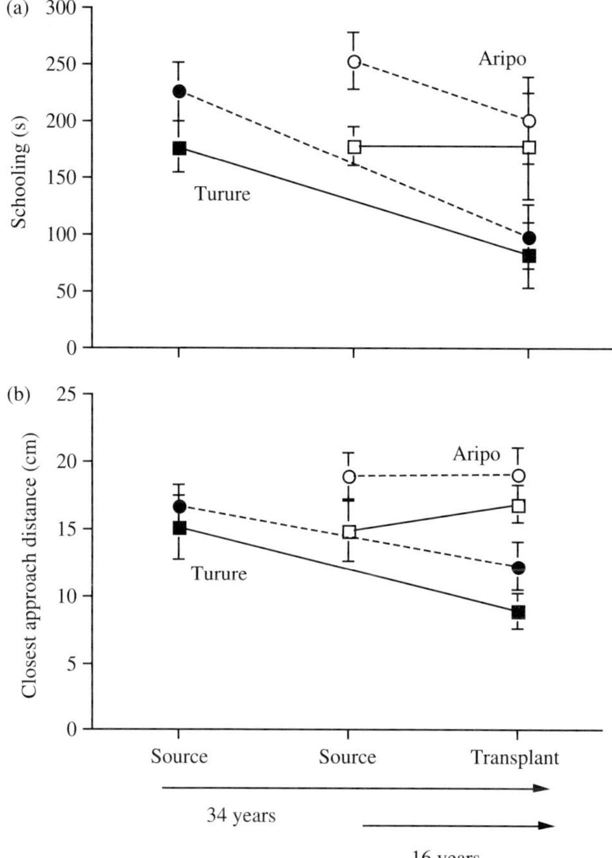

Fig. 3.4 Evolution of anti-predator behaviour in guppies. This figure summarizes the consequences for two components of anti-predator behaviour of a shift from a high-predation habitat to a low-predation one. The Turure transplant was undertaken by Caryl Haskins in 1957, the Aripo transplant by John Endler in 1976: see Table 3.2 for details. A period of 34 and 16 years, respectively had elapsed between the transplant and the behavioural assay. Fish were collected from the wild and then bred, raised, and tested under standard conditions in the laboratory. The graph compares the behaviour of the descendants of the transplanted fish with a sample from the source population. Schooling tendency (a) is the number of seconds that the focal individual spent in the proximity of a school of guppies of the same sex. Closest approach distance (b) is the minimum distance between the focal guppy and a realistic *Crenicichla* model during inspection. Females are represented by round symbols (and dashed lines) males by square ones. Solid symbols denote fish from the Turure transplant. In all cases means and 95% confidence intervals are presented. Fig. (a) is redrawn from Magurran (1998).

survivors was then recorded. The results confirmed, as might be expected from the earlier studies, that guppies that co-occur with pike cichlids are much better at avoiding them. The escape ability of fish from the introduced populations was thus predicted by their geography rather than by their ancestry. But the most important finding was that the qualitative difference between introduced and ancestral populations was replicated in the F_2 generation. This result supports the hypothesis that population differences in escape ability are underpinned by genetic effects and confirms that evolution in this trait can occur rapidly (within 15–20 years, or approximately 26–36 generations).

O'Steen and co-workers' experiment raises a series of interesting issues. The investigators found that the difference in behaviour between the 'introduced pairs' was less marked than between the 'natural pairs'. They also discovered that the differential survival of guppies derived from a high-predation locality was moderated in the F_2 generation. These results suggest that the behaviour of the transplanted fish has not yet converged on the optimal for the habitat, and importantly, that the magnitude of the response is determined by phenotypic effects, such as learning. Additional support for these ideas can be gleaned from an earlier study. In 1992 (that is 5 years before O'Steen and co-workers' main set of experiments) my colleagues and I (Magurran and Seghers 1994*b*; Magurran *et al.* 1995) examined the anti-predator behaviour of fish in the Aripo River introduction site (Fig. 3.5). Two further sites, the naturally low predation Upper Aripo, and the Middle Aripo, to which *Crenicichla* had been introduced in 1981, completed the survey. We found that the predator inspection behaviour of the wild fish in both introduced sites was consistent with what would be expected for this type of environment (Fig. 3.5). The same was true for schooling behaviour. For example, guppies in the new low-risk locality spent less time schooling, and were less wary of potential predators than the ancestral population. However, when we raised fish under standard conditions in the laboratory these differences diminished and guppies from both the ancestral and introduction Aripo sites behaved in a similar manner (Fig. 3.4(a) and (b)). The fact that O'Steen *et al.* found a difference in survival between fish in the introduction site and its high-predation founder population implies that genetic differences were strengthened during the interval between the studies—a result consistent with the observation that a significant change in behaviour had been previously observed in the longer running Turure introduction (Fig. 3.4). As the Magurran *et al.* and O'Steen experiments measured different types of anti-predator behaviour I cannot be certain of this interpretation. Nonetheless, both investigations show how important environmental effects are in shaping behaviour.

One possibility is that a change in predation risk selects first on phenotypic plasticity. This makes sense when we consider that predation risk varies substantially over space and time. Flexible behaviour allows an individual to respond to local conditions without being encumbered by a defensive system that over-reacts to non-threatening stimuli, and thus wastes time and energy, or under-reacts and puts the animal in danger of death. There are two ways in which this plasticity could be modified. The threshold required for long-term modification of behaviour could be lowered or heightened depending upon the severity of local risk, or the extent to

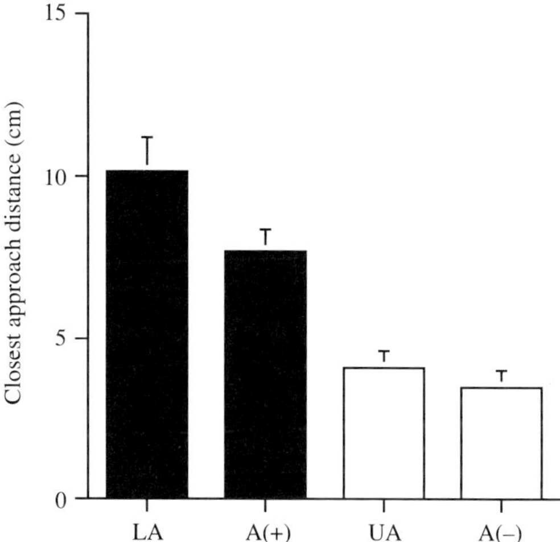

Fig. 3.5 Mean (+95% confidence interval) closest approach distance during inspection of a model *Crenicichla* by wild guppies in the Aripo River. Black columns represent localities where *Crenicichla* currently co-occur with guppies. Open columns show *Rivulus* sites. Guppies in two of the sites originally experienced the obverse predation regime—*Crenicichla* were introduced into A(+), (the Middle Aripo) in 1981 while guppies were transplanted from the high-risk LA (Lower Aripo) to A(−) (Aripo tributary) in 1976. UA is the low-risk Upper Aripo (Naranjo). These data were gathered in 1992 during the investigation described by Magurran and Seghers (1994*b*).

which a behaviour pattern is altered following a priming event might change. Some work I (Magurran 1990*b*) did with European minnows, *P. phoxinus*, provides support for the latter mechanism but does not rule out the former. I discovered that an 'attack' by a model pike (*E. lucius*) on 2-month-old fish translated into enhanced predator-evasion abilities in adult fish tested almost 2 years later. All fish showed a capacity for learning but the improvement in anti-predator behaviour was greater in fish that had coexisted with predators (particularly the pike *E. lucius*) over historical times. This finding recalls Konrad Lorenz's (1966) 'innate school-marm'—the notion that animals are predisposed to learn certain tasks. Felicity Huntingford and Peter Wright have shown that newly hatched sticklebacks, *Gasterosteus aculeatus*, from high-risk populations, react more vigorously to their father's retrieval attempts when they are in the nest, and that these interactions strengthen their anti-predator responses (Huntingford and Wright 1993). Older sticklebacks also exhibit a population-specific ability to learn about danger (Huntingford and Wright 1989; Wright and Huntingford 1992).

These studies jointly point towards an effect of predation risk on learning. They additionally show that the baseline behaviour of individuals, raised in a common environment and in the absence of the triggering stimulus, is predicted by origin. This implies that some components of the anti-predator repertoire are hard-wired. The consistent pattern is that the same qualitative elements are found in the behavioural repertoire of all populations, but that the quantitative expression of those elements varies. For example, all guppies school, but the amount of time an individual spends schooling, and the structure of the school it joins, is influenced by the level of predation risk its ancestors have experienced.

With the exception of the 'common garden' experiments described above there have been few attempts to investigate the genetic architecture of predator-evasion traits in guppies. Paxton (1994) used a hybridization approach and made crosses and back-crosses between a high-predation and low-predation population (Lower and Upper Aripo, respectively). He discovered that the shoaling tendency of individual females increased in line with the proportion of genetic material derived from the high-predation stock. Paxton estimated the upper limit for narrow-sense heritability as 0.25. (Narrow-sense heritability is the proportion of total phenotypic variation due to the additive effect of the genes and as such the variation that natural selection acts upon). Wright *et al.* (2003) used a different approach—a North Carolina II breeding design (which uses multiple fathers to produce half-sib progeny within a family)—to estimate shoaling tendency of zebrafish, *Danio rerio*. They obtained a heritability estimate for shoaling tendency of 0.23 ± 0.25. When within fish variance was excluded from the estimate of total phenotypic variance, shoaling heritability (again narrow-sense) increased to 0.40 ± 0.41. These estimates are in line with John Endler's expectation (cited in Magurran 1999) that the heritability of behavioural traits in guppies and similar fish will be in the order of 0.3–0.5 or less. Wright *et al.* (2003) further confirmed a genetic basis for boldness in zebrafish while Paxton (1994) found that the genetic background of the fish influenced various aspects of their inspection behaviour. The use of the QTL approach to map various traits in guppies has been impeded by the absence of a comprehensive suite of molecular markers. However, a new initiative to construct a guppy 'map' (see Chapter 8) will open up important opportunities for researchers investigating the genetic basis of ecologically significant behaviour.

3.4 Kinship, familiarity, and predator avoidance

The congregation of guppies into shoals and schools begs the question of whether these are random assemblages of individuals or whether particular fish associate more than would be expected by chance. Researchers have long been interested in the possibility that fish schools might be composed of related individuals since this would provide a plausible explanation for the maintenance of apparently altruistic behaviour, such as the production of alarm pheromones (Smith 1992) and predator inspection. A simple mechanism that can promote kin groupings is the association of related individuals from birth or hatching. A common sight in Trinidad, for instance, is a school of young *Crenicichla* being shepherded by both parents. I am not aware that the kinship of these schools has been formally measured but in the light of the monogamy and biparental care shown by pike cichlids I would expect relatedness to be high. However, genetic analyses of adult schools of various species have uncovered scant evidence of kin groupings (Ferguson and Noakes 1980; Dowling and

Moore 1986; Griffiths 2003), even in those fish that appear able to discriminate relatives in the laboratory. The threespine stickleback, *G. aculeatus*, is a good example (Van Havre and Fitzgerald 1988; FitzGerald and Morrisette 1992; Peuhkuri and Seppa 1998). Some life-history traits, such as broadcast spawning, mitigate against the formation of kin groups (Gerlach *et al.* 2001). In other cases the failure to detect kin associations may be attributable to methodological biases, such as the difficulty of collecting entire schools (Avise and Shapiro 1986; Dowling and Moore 1986; Naish *et al.* 1993; Herbinger *et al.* 1997), or the use of molecular tools, such as allozymes, that provide low resolution of kin relationships (Avise and Shapiro 1986; Peuhkuri and Seppa 1998).

On the face of it, guppy schools should be a good place to search for kin associations. Females produce small broods of live young. These individuals are full or half-siblings (Becher and Magurran 2004) and can school from birth (Magurran and Seghers 1990*b*). There is no ontogenetic shift in schooling tendency that might cause erstwhile schooling partners to pursue a solitary existence. Females (more than males) are known to associate with the same school mates over time (Griffiths and Magurran 1998) and show considerable site fidelity (Haskins *et al.* 1961; Magurran *et al.* 1994; Croft *et al.* 2003*a*, *b*). Moreover, the development of microsatellites for guppies (e.g. Becher *et al.* 2002; Shikano and Taniguchi 2003) means that relatedness can be assessed with precision. Russell *et al.* (2004) collected 25 complete adult shoals of guppies from two high-predation localities in Trinidad—the Quare (in the Oropouche drainage) and the Lower Tacarigua (in the Caroni drainage). Shoals were observed prior to capture so that we could be sure they were captured in their entirety. Fish were genotyped using seven hypervariable microsatellite loci. We expected to find that female (but not male) members of these shoals would be more closely related than the average for the population. However, this hypothesis was not supported. In contrast to a previous study that had used less precise markers (Magurran *et al.* 1995) we (Russell *et al.* 2004) found that neither females nor males schooled with kin and that the relatedness of shoal members was not significantly different from that predicted by chance. These results suggest high turnover in shoal composition and strengthen the conclusion that schools of adult fish are rarely (if ever) comprised of kin. This does not preclude the possibility that shoals may possess sub-structure and that on occasion individuals (and in guppies this probably means females) will preferentially be found with kin (Russell *et al.* 2004). Careful observation of individual schooling choices in the wild, followed up by genetic analysis, would be needed to resolve this. There have been two laboratory studies of kin discrimination by guppies. Both of these showed that familiarity was more important than kinship in determining shoaling preferences (Warburton and Lees 1996; Griffiths and Magurran 1999). Indeed, familiarity effects may have confounded a number of studies that set out to demonstrate kin-based associations (Griffiths 2003 table 1) for although familiarity may be used by some species as a surrogate of kinship (Mann *et al.* 2003), investigations that do not control for it (e.g. by rearing unrelated individuals together) do not prove kin recognition.

The absence of kin-based shoals does not mean that kinship is unimportant to fish. There are many studies indicating that fish have the ability to distinguish relatives from non-relatives (see Griffiths 2003 for a review). Interestingly, fish may actively avoid relatives. This has been demonstrated in territorial animals and seems to arise when competition for resources is most intense between closely related individuals. Griffiths and Armstrong (2001), for example, found kin dispersal rather than kin association in wild Atlantic salmon, *Salmo salar*, while Greenberg *et al.* (2002) showed that a reduction in mean relatedness increased the growth rates of juvenile brown trout, *Salmo trutta*, in outdoor enclosures.

One by-product of the work on inspection behaviour was the discovery that fish can choose among individuals on the basis of prior experience. Manfred Milinski and his colleagues (Milinski *et al.* 1990*a*, *b*) found that inspecting sticklebacks are able to build up 'trust' and share the risk involved in approaching a potentially dangerous predator. Dugatkin and Alfieri (1991*a*) similarly showed that guppies can differentiate individuals on the basis of their behaviour and in future encounters will choose to be near fish that previously inspected most assiduously. Recent work has confirmed that fish of many species preferentially associate with familiar conspecifics (Griffiths 2003 table 2).

The ability to recognize particular individuals has a number of important advantages in addition to selection of inspection partner. Chivers *et al.* (1995) discovered that schools of fathead minnows, *Pimephales promelas*, execute anti-predator manoeuvres more effectively when they consist of familiar rather than unfamiliar shoalmates. Metcalfe and Thomson (1995) showed that European minnows, *P. phoxinus*, could identify individuals that were less effective foraging competitors, and associated with them preferentially. Kelley *et al.* (1999) demonstrated that male guppies preferred to court unfamiliar females (see also Chapter 4).

Investigations of guppies have helped uncover the degrees of freedom under which familiarity operates. Magurran *et al.* (1994) confirmed that guppies are able to discriminate familiar from unfamiliar shoaling partners. (This study also revealed that there is no preferential association with individuals from the same population when levels of familiarity are controlled). Griffiths and Magurran (1997*a*) subsequently found that female guppies had to be together for around 12 days before familiar fish were distinguished from unfamiliar ones. Other studies (summarized by Griffiths 2003) provide support for the idea that familiarity is acquired over periods of a few days to a few weeks. Although wild male guppies do not appear to use familiarity to modulate their shoaling behaviour (Griffiths and Magurran 1998; Godin *et al.* 2003), they will preferentially associate with same-sex conspecifics (Croft *et al.* 2004*a*) when held together for 12 days—the time period over which female guppies learn to distinguish familiar individuals (Griffiths and Magurran 1997*a*). The sex difference in the expression of shoaling preferences for individuals is probably rooted in the mating system of the guppy. Males potentially increase their reproductive success by moving among shoals in search of new mating partners (Kelley *et al.* 1999). However, familiarity is constrained, not just by the length of time that fish have been able to associate, but also by the number of individuals with which they can interact. In an investigation involving wild fish in Trinidad's Upper Tunapuna River, where guppies are confined to a series of isolated pools during the dry season, Griffiths and Magurran (1997*b*) discovered that shoaling preferences for familiar females are inversely related to pool population size. Once the number of females per pool exceeded about 50 a focal female was no more likely to associate with her pool mates than with random females drawn from the Upper Tunapuna. This suggests either that fish do not have the cognitive ability to distinguish individuals above a threshold number, or that the advantages of discrimination follow the law of diminishing returns. For instance, the safety in numbers advantages of a larger school size might cancel out the benefits of associating with particular individuals. There is also accumulating evidence that the extent to which shoaling decisions are guided by familiarity varies over space and time. In contrast to Griffiths and Magurran (1998), who worked on a high-predation (and 'high shoaling') population, Godin *et al.* (2003) found that low-predation female guppies did not preferentially associate with familiar same-sex individuals. This may be attributable to the fact that fish that are in no immediate danger of predation, and that gain no foraging advantages from associating with familiar individuals, draw fewer benefits from shoaling with fish that are known to them (Godin *et al.* 2003). But it is also of interest that female guppies in the same locality (the Upper Tunapuna) showed different tendencies to shoal with familiar individuals when tested on different occasions (Griffiths and Magurran 1997*b*; Godin *et al.* 2003). Temporal variation in the ecology and social structure of guppy populations influences individual behavioural decisions in ways we have only just begun to understand.

3.5 Populations and learning

It is well known that fish, along with other animals, can improve their anti-predator skills through learning (Brown 2003; Kelley and Magurran 2003*b*). Guppy populations provide many opportunities to test ideas in this rapidly expanding field. There is some evidence that experience when very young improves predator-evasion behaviour. In one of the first studies to examine the ways in which different types of experience might lead to improved anti-predator behaviour, Goodey and Liley (1986) exposed guppies in the first 48 h of life to a variety of cues: chasing by adult guppies, visual cues, or chemical cues, of predators, or guppy chemical cues. A final group were isolated from all predator and conspecific cues. Goodey and Liley's study suggests that fish that have been chased as juveniles are better able to withstand an attack by a predator (Fig. 3.6). The results are, however, complicated by the fact that there may have been selection (through cannibalism) against less proficient evaders in the 'chasing' treatment. Jennifer Kelley (2002) repeated part of the experiment with a more careful handling control. Newborn guppies were assigned to one of three treatments: repeated exposure to chasing adult conspecifics; repeated exposure to the observation arena; or no handling. Various aspects of inspection behaviour towards a live *Rivulus* were tested in 6-week-old fish. Kelley found no clear differences in inspection behaviour among treatments but did observe that behaviour varied with

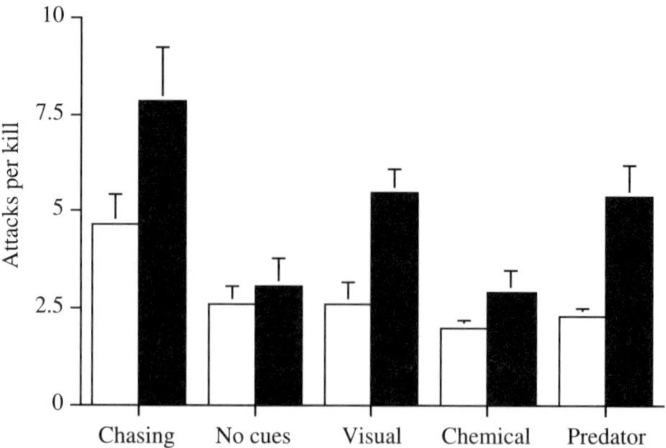

Fig. 3.6 Goodey and Liley (1986) exposed young (<48-hour-old) guppies to a variety of treatments. Fish either experienced 'chasing' from adult conspecifics, 'no cues' from any fish, received 'visual' cues when they watched other juveniles being chased, or 'chemical' cues in the form of water from a tank of chasing adults, or encountered 'predator' cues—water from tanks containing cichlid and cyprinodontid predators. After maturity, their escape responses were tested in encounters with *Cichlasoma nigrofasciatum* (in the case of males) and *Crenicichla* sp. or *Pseudotropheus* sp. (in the case of females). In both sexes fish that had been chased as juveniles tended to survive more attacks before being killed. Males are denoted by open bars, females by filled ones. Mean values (+ s.e.) are shown.

body size (with larger fish showing more attack-cone avoidance and approaching the predator more closely). The consequences of early experience, and the potency of different sorts of cue—as well as the interaction with fish origin (the severity of the predation regime of population from which the fish are drawn)—remain to be elucidated in detail. This is not simply a deficiency in the guppy literature but a general omission in the field. As Kelley and Magurran (2003*b*) observe, no studies on fish have specifically examined ontogenetic changes in the ability to learn about predators. It is also possible that events prior to birth may influence subsequent behaviour. Chemical information would be one way of achieving this. Mirza *et al.* (2001) found that brook charr, *Salvelinus fontinalis*, eggs hatched earlier in the presence of chemical cues from predatory sculpins, *Cottus cognatus*. It might be interesting to ask whether embryos can make use of chemical information from the mother, or even if the number or sex ratio of embryos in the uterus can influence subsequent behaviour. Advances in developmental biology, including the ability to raise guppy embryos *in vitro* (C. Dreyer, personal communication), present new opportunities to investigate learning skills in very young fish.

Social learning, on the other hand, has received considerable attention. Social learning can be defined as learning that makes use of socially provided information (Brown and Laland 2003). It embraces 'local enhancement'—where naïve individuals are attracted to presence of other fish, and learn something as a consequence, 'social facilitation'—where engaging in the behaviour initiated by other individuals induces learning, and 'observational conditioning'—where conditioning to a particular stimulus is achieved when fish copy the behaviour of individuals that have already been conditioned (Brown and Laland 2003). A series of cleverly designed experiments by Kevin Laland and his colleagues (see, for example, Laland and Williams 1997; Lachlan *et al.* 1998; Laland and Reader 1999) has demonstrated that information about foraging sites can be obtained through social learning. Guppies can also improve their anti-predator behaviour through social learning (Sugita 1980). Brown and Laland (2003) found that naïve fish copied the escape route (through a trawl net) used by trained demonstrators, and that this experience increased their speed of escape in subsequent trials, even though they did not necessarily continue to use the demonstrated route. Kelley *et al.* (2003) asked whether guppies from a low-predation population (Tunapuna) could improve their anti-predator behaviour through interaction with fish from a high-predation population (Tacarigua). Guppies trained with high-predation—but not low-predation—demonstrators significantly increased their schooling time and inspected a realistic predator model from further away. Interestingly, naïve fish that associated with the experienced demonstrators in the absence of a predation threat showed no enhancement of their anti-predator behaviour. This suggests that an overt anti-predator response, rather than the higher baseline schooling behaviour of the demonstrator population, was crucial in the learning. Although high-predation guppies and low-predation guppies will only infrequently interact in the wild the same process could enable less experienced younger fish to become more proficient at avoiding predators.

Social learning is also implicated in acquired recognition of alarm cues. Suboski *et al.* (1990) discovered that naïve zebra fish (*D. rerio*) would respond to an artificial odour if paired with individuals that had been conditioned to respond to it. Mathis *et al.* (1996) confirmed that social learning enabled naïve fish (fathead minnows) to recognize novel predator cues. Given the demonstrated ability of guppies to recognize chemical alarm cues (Nordell 1998; Brown and Godin 1999), and the importance of olfactory information in the aquatic environment, the species offers profitable opportunities to investigate geographic variation in learning (including social learning) of different types of cue. I would predict, for instance, that olfactory cues would be accorded more weight in relatively still and turbid lowland rivers than in the fast flowing and usually clear mountain streams.

3.6 Ontogenetic shifts in behaviour and morphology

Viviparity produces guppies that are independent at birth. There have been relatively few investigations of the behaviour of newborn guppies. Magurran and Seghers (1990*b*) discovered that guppies could school from birth. This ability is clearly advantageous since small individuals are vulnerable to cannibalism from conspecifics and predation by invertebrates and other juvenile fish species. Newborn guppies also undertake predator inspections and can execute coordinated evasion tactics such as the flash expansion manoeuvre (pers. obs). Evans and Magurran (2000) found that females that had mated

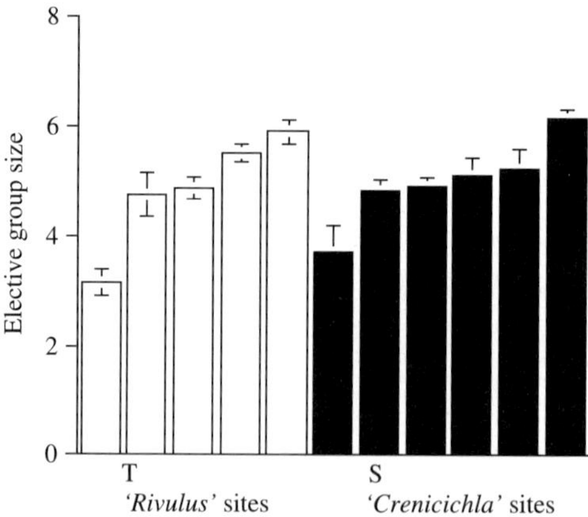

Fig. 3.7 Schooling behaviour of newborn guppies. Newborn fish (<48-hour-old) were allowed to swim freely in groups of eight in an open, circular arena (41 cm diameter). The tendency of fish to congregate in a school was measured by recording the elective group size (that is the distribution of fish among schools) every 30 s for 10 min. Schools were defined as groups of fish in which no individual was more that 5 body lengths from its nearest neighbour. Mean maximum group size (± s.e.) is presented. No individual was tested more than once. In total 11 populations were examined. Although there was an overall significant difference in schooling behaviour (one-way ANOVA $F_{10,130} = 21.79$, $P < 0.001$), this was not related in any obvious way to predation regime. ('*Rivulus*' sites are denoted by open bars, '*Crenicichla*' sites by closed ones). The Upper Turure (Haskins transplant) and Guanapo (Haskins source) are represented by T and S, respectively. Other populations, from left to right, are Upper Paria, Upper Yarra, Upper Tunapuna, and Upper Aripo (all '*Rivulus*' sites) and Lower Oropouche, Tranquille, Lower Quare, Lower Turure and Lower Aripo.

with more than one male produced broods with improved schooling and predator-evasion behaviour. However, in an experiment that utilized artificial insemination Evans *et al.* (2004*b*) found that the escape response, but not the schooling behaviour, of newborn offspring was influenced by sire attractiveness. This suggests that sire genotype affects some aspects of offspring performance but that maternal affects may influence others. Offspring size—which may in part be under female control—could have an effect on schooling behaviour (Fuiman and Magurran 1994). It turns out that offspring in the broods produced by the multiply mated females in Evans and Magurran's (2000) experiment were slightly larger (as well as more in number) than those sired by a single father (Ojanguren *et al.* 2005). These ideas are discussed further in Chapter 4. The schooling tendency of newborn guppies varies among populations in a manner that is difficult to interpret (Fig. 3.7). Upper Aripo females produce some of the largest juveniles so it is possible that variation in offspring size accounts for some of this pattern. Risk to newborns almost certainly varies geographically. Cannibalism seems to be greater in some populations than others and juvenile guppies appear to vary in cover seeking behaviour. I am unaware of any studies that have formally investigated this. The shifts in shoaling and other anti-predator responses in relation to ontogeny, particularly

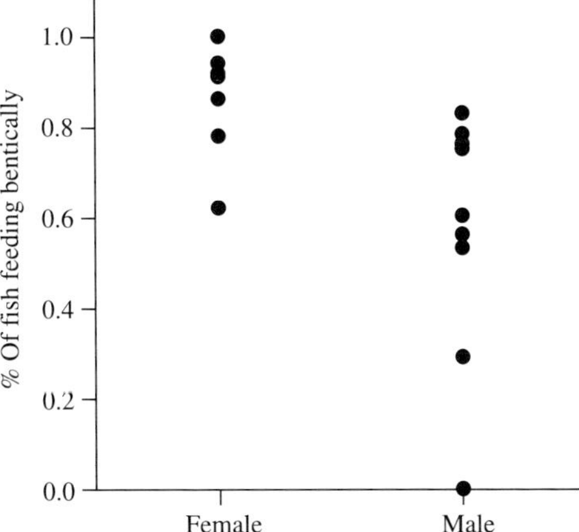

Fig. 3.8 Female guppies are more likely than males to feed on the bottom of a river or pool. This graph shows the proportion of adult individuals of each sex feeding on the benthos. Data points represent averages for 10 separate pools in the Upper Tunapuna River. After Magurran (1998).

as sexual maturity is achieved, would make an interesting study. Predation risk will of course vary with life-stage.

One topic that has been investigated is the potential for induced morphological diversity. Robinson and Wilson (1995) took juvenile (<1-week-old) guppies and randomly assigned them to five feeding treatments. Fish in four of the treatments were offered food in different locations, such as the bottom or surface of the tank. In the final treatment the food source was rotated over 4 days. Data from other species show that phenotypic plasticity can account for functional diversification in feeding modes in fish (Meyer 1987, 1989; Robinson and Wilson 1994). In Robinson and Wilson's guppy experiment males that experienced the floating food regime had shallower and longer bodies, longer skulls, and longer paired fins than those in other treatments. Females in contrast exhibited no morphological diversification associated with feeding regime. It is possible that females, perhaps as a result of their live-bearing habit, show reduced phenotyoic plasticity. Furthermore, it is worth noting that in the wild males and females may adopt feeding niches that are consistent with their dimorphism in morphology. Males with their fusiform body shape can be found foraging in mid-water whereas heavier bodied females are more likely to feed on the bottom of the river or pool (Magurran 1998) (see Fig. 3.8): These sex differences in feeding behaviour may constrain the evolution of phenotypic plasticity, particularly for females, in this species.

3.7 Differences between the sexes in response to predation

George Williams (1964*b*, p. 368) was one of the first investigators to observe that the schooling behaviour of male guppies is much less pronounced than that of females.

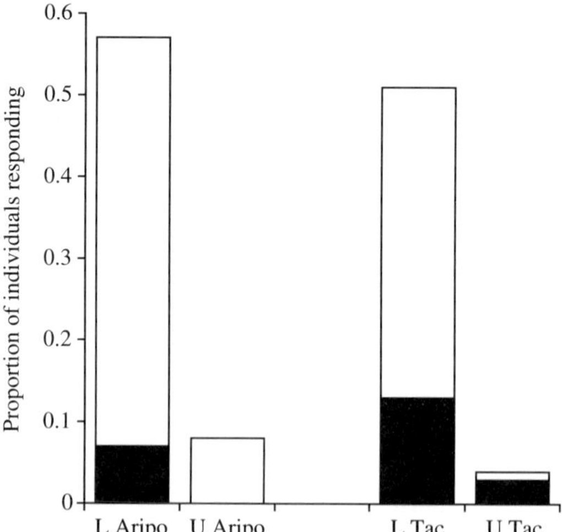

Fig. 3.9 Females are more wary than males, particularly in high-predation localities. This figure shows the proportion of wild females (open bar) and males (solid bar) displaying anti-predator responses during a focal 3 min period. L Aripo and U Aripo represent high- and low-predation localities in the Aripo River, L Tac, and U Tac the equivalent in the Tacarigua River. After Magurran and Seghers 1994*c*. Fig. 3.4 provides further insights into the weaker anti-predator responses of males.

Their behaviour was, he commented, 'obviously intermittent and of low intensity'. Although Williams was working with 'pet-store' guppies the same pattern is clearly apparent in the wild. Indeed, one of the most striking behavioural contrasts in guppies is the difference between the sexes. As Fig. 3.9 reveals, females are consistently more wary than males. They spend more time schooling, and engage in more inspection behaviour but maintain a greater distance between themselves and the predator during an approach. Croft *et al.* (2004*b*) report that under high-predation risk sexual segregation occurs because the more vulnerable males congregate in shallow, marginal habitats. Indeed, there are multiple aspects of guppy behaviour, particularly those related to predator defence, that vary between the sexes (Magurran and Macías-Garcia 2000 and see Table 3.3). The greater caution of females in the face of predation risk can be seen in other species too. For example, females of the viviparous fish *Girardinichthys multiradiatus* approach predators less often than dominant males (Macías-García *et al.* 1994), and are less frequently consumed by predators than males (Macías-Garcia *et al.* 1998). These sex differences in behaviour are rooted in differences in reproductive potential. In the case of females reproductive success is correlated with longevity. As mating partners are rarely in short supply and females can store sperm, additional progeny will be produced with each successive brood cycle without the need to mate repeatedly. (This is not to say that brood quality might not also be enhanced through polyandry—an issue I explore in the next chapter).

Table 3.3 Sex differences in guppy behaviour

Trait	Male	Female	
Coloration	Brightly coloured: polymorphic	Uniform cryptic beige	(Houde 1997)
Size	Smaller (13–19 mm L_S)	Larger (18–24 mm L_S)	(Reznick and Miles 1989)
Morphology	Fusiform shape—for pursing females	Deeper-bodied	(Robinson and Wilson 1995)
Fins	Ornamented caudal and dorsal fins: gonopodium	Unornamented fins: normal anal fins	(Houde 1997)
Foraging effort	15–30% of time budget in the wild	45–75% of time budget in the wild	(Dussault and Kramer 1981; Magurran and Seghers 1994*c*)
Schooling tendency	Weaker schooling tendency—typically about 20% lower than females	Stronger schooling tendency	(Magurran *et al.* 1992; Magurran 1998)
Schooling partner	Frequent switches between schools	May remain in established schools of familiar individuals	(Griffiths and Magurran 1998)
Anti-predator responses	Reduced	Elevated—on average four times more likely to respond	(Magurran and Seghers 1994*c*)
Vigilance	Less vigilant—initiate predator inspection on *c*.20% of occasions	More vigilant—initiate predator inspection on *c*.80% of occasions	(Magurran and Nowak 1991)
Predator inspection	Less cautious—average minimum distance from predator during inspection 15 cm	More cautious—average minimum distance from predator during inspection 20 cm	(Magurran *et al.* 1992)
Sexual segregation	Males congregate in shallow, marginal habitats under high-predation risk	Females in deeper water	(Croft *et al.* 2004*b*)
Risk of mortality due to predation	Higher	Lower	(Seghers 1973; Rodd and Reznick 1997)

Source: (Magurran and Macías-Garcia 2000)

Males, on the other hand, while having the possibility of siring many offspring may fail to father any progeny at all, irrespective of how long they live. These indirect consequences of risk are investigated in the following chapter.

3.8 Conclusions

Trinidadian guppies, in part, because they have been so thoroughly investigated, but also because of the well-documented geography of risk, exemplify the behavioural and morphological responses that prey show to predators. Transplant experiments have confirmed that these traits evolve as a result of a shift in predation risk, though there are interesting and as yet, incompletely understood differences in the length of time needed for heritable differences to become established. In part, this may be due to the tendency to treat 'predation risk' as a unidirectional and consistent force when in reality it will be mediated by the population dynamics of both predators and prey, by assemblage structure and by changes in the habitat. Another important factor is the contribution that learning makes to the execution of anti-predator responses, and the extent to which it is itself a target of selection.

Guppies have provided some of the best tests of key theories in evolutionary biology, such as the potential for cooperation during predator evasion and the role of kinship and familiarity in decision-making. Although a number of investigators have attempted to explore these questions in the wild, as well as in a controlled laboratory environment, much remains to be learnt about the social dynamics of these fish under natural conditions.

4

Reproduction

Trinidadian guppies have become one of the classic examples of sexual selection. Not only are the bright and varied colour patterns of wild males aesthetically pleasing, but they may also signal fathers that are in better condition or who are genetically superior (Reynolds and Gross 1992; Evans *et al.* 2003*b*, 2004*b*). It might seem obvious from the perspective of the early twenty-first century that male colour is a sexual signal and that females accrue direct or indirect benefits from using it in their mating decisions. However, it was not always so and even a century after Darwin (1859, 1871) explained how secondary sexual characters might improve an individual's mating success, Haskins *et al.* (1961, p. 387) commented that they had 'not been able to pin down any firm evidence of male selection by the female' while Liley (1966, p. 183) made the rather tentative statement that 'sexual selection may operate in these species' (*Poecilia reticulata*, *Poecilia parae*, *Poecilia picta* and *Poecilia vivipara*). Yet, guppy investigators were by no means remiss in this and indeed were ahead of their time in recognizing the potential for sexual selection to operate. As Harvey and Bradbury (1991) make clear, despite some significant exceptions, including the writing of R. A. Fisher and Julian Huxley, Darwin's ideas were neglected for many decades. A handful of empirical studies provided insights into aspects of sexual selection. For example, Bateman (1948) showed that the fitness of *Drosophila* males was more variable than females while Tebb and Thoday (1956), also using fruit flies, found that differences among females in their preferences for male characters resulted in sexual selection. Peter O'Donald (O'Donald 1967, 1973, 1977, 1980) revisited Fisher's work (Fisher 1914, 1930) and developed mathematical models that showed, among other things, how female preferences and male characters might become associated. But it was not until the 1970s, when the tsunami of behavioural ecology swept up laboratory, field, and theoretical biologists in its path, that sexual selection gained anything like the prominence it receives today.

As sexual selection in guppies, particularly in the form of female choice, is already very well covered in Anne Houde's excellent monograph *Sex, color, and mate choice in guppies* (Houde 1997) it will not be revisited in detail here. Instead, the chapter will develop themes initiated by the early workers and show how recent investigations have shed light on some of the important questions they raised. New molecular techniques provide unrivalled opportunities to answer questions that some of the earlier researchers could only speculate about. One of the most exciting advances addresses the role of post-copulatory mechanisms. Interestingly though, despite the methodological advances of the last decade or so, some early problems are not yet resolved.

An example is sneaky mating behaviour. Despite many studies that attest to its prominence in natural populations it is still uncertain to what extent (if indeed at all) the tactic translates into paternity. The chapter explores these issues. The initial focus, which makes the link with the preceding chapters, is on the indirect consequences for mating behaviour of variation in risk. However, the chapter begins with a short review of reproductive biology and behaviour.

4.1 Reproductive biology and behaviour

As noted in the introduction male guppies, along with other poeciliids, have a modified anal fin, known as a gonopodium, that is used as an intromittent organ. The third, fourth, and fifth rays of the gonopodium (Fig. 4.1) are modified to form a channel, down which the sperm bundles or spermatozeugmata (Philippi 1908) pass into the female. The small hook at the tip of the gonopodium is much less pronounced in guppies than in some other species of poeciliid. For instance, the nondescript-looking livebearer *Tomeurus gracilis*, a species without even a common English name, has an elaborate gonopodium that is festooned with structures that have been described as resembling sickles, grappling hooks, and ice tongs (Rosen and Bailey 1963; Constanz 1989). The gonopodium also has a 'hood', which appears to have a sensory function (Clark and Aronson 1951). Although the hood is not necessary for successful insemination, Houde (1997) points out that its development accompanies the onset of sexual maturity and that it is only when the hood grows beyond the end of the gonopodium that males seem able to fertilize females. The male whose gonopodium is depicted in Fig. 4.1 was therefore probably just on the threshold of maturity. Sperm bundles are thought to comprise around 22,000 sperm each (Fig. 4.2) (Billard 1969). The size of

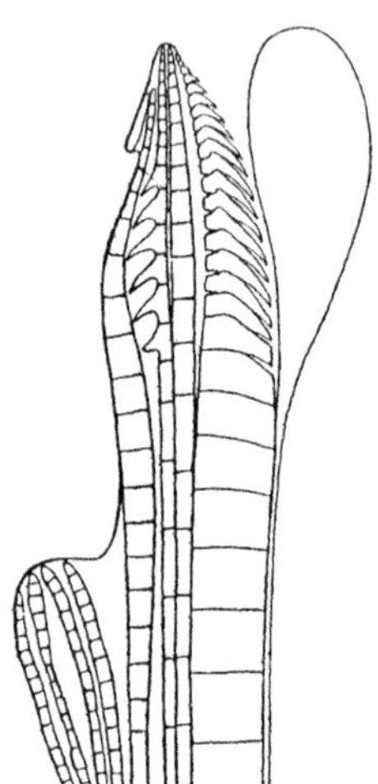

Fig. 4.1 Guppy gonopodium (Regan 1913).

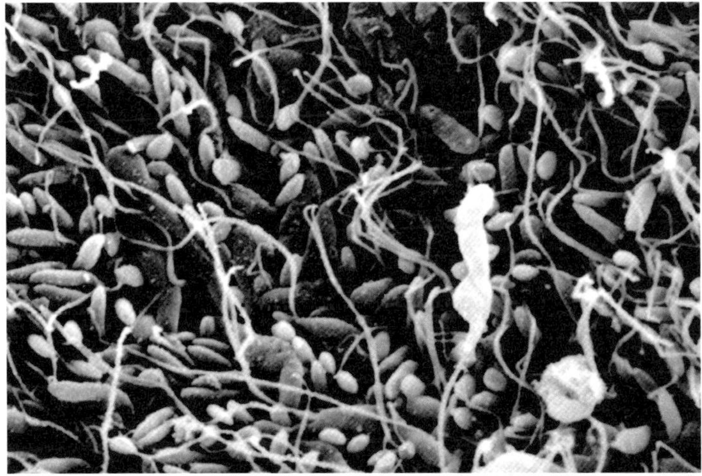

Fig. 4.2 Guppy sperm in testis (photograph by I. M. Matthews).

natural ejaculates is in the order of ~500,000 sperm (Evans *et al.* 2003*b*) and males have been estimated to produce ~750,000 sperm per day (Billard 1986). Bozynski and Liley (2003) found that female presence significantly increased male sperm reserves ('strippable sperm'). Males that were able to observe a gravid female for 1 week had an average sperm count of 3.4×10^6. This contrasted with 1.3×10^6 in the control males where no female was present. With the exception of the sperm bundles, which appear to be fairly uniform in size, there are large confidence intervals around these estimates.

Male guppies have two means of securing a mating. They can either display and seek a consensual mating with a female, or alternatively engage in sneaky mating (sometimes referred to as gonopodial thrusting) without the female's consent. All males employ both tactics, sometimes in quick succession, although there are marked individual differences in both the frequency and relative usage of the two behaviours (Magurran and Seghers 1990*c*) as well as differences among populations (Luyten and Liley 1985). Mating behaviour has been comprehensively catalogued by Baerends *et al.* (1955) and by Liley (1966). A courting male will typically follow a female and display when he comes within her field of view. During the sigmoid display the male's body assumes an S shape and the dorsal and tail fins are either extended or folded. The whole body quivers and courting males become noticeably more colourful. Indeed, the change, particularly in the black colour spots, is sometimes so marked that it can be difficult to reconcile the appearance of a courting male with his non-courting self. Houde (1997) notes that individual males vary in the expression of their colour patterns. The duration of a display—which can last for several seconds—depends on the male's motivation, the female's interest, and the presence of competitors. It is not unusual for a male to abandon a display and switch to gonopodial thrusting if rival males start to pursue 'his' female. The relationship between display rate and female choice remains a matter of some debate. Although there are a number of studies indicating that females are attracted to higher displaying males (Farr 1980*b*; Stoner and Breden 1988; Kodric-Brown 1993) the evidence that display rate translates into paternity, independently of variation in colour or other traits, is weak. Houde (1988*a*) uncovered a negative correlation between mating success and male display rate. She (Houde 1997) suggests that it is the way in which males use their display rather than the frequency with which they employ it that is important. Houde's idea are consistent with a study (Becher and Magurran 2004) that used molecular markers to assign paternities and found no association between male behaviour and mating success. Nonetheless, males with relatively high display rates gained more paternity in an investigation that examined sperm competition between the first and second male to mate consensually with a virgin female (Evans and Magurran 2001). It is also possible that tactile and olfactory stimuli are being appraised as well as the visual ones that appear most prominent to us as human observers.

If a female is receptive—which means that she is either a virgin or has recently given birth to a brood (Liley 1966), or has been deprived of access to males for many weeks—she may approach the male with a 'glide' response. Females typically 'arch' their bodies prior to copulation during which the male and female will 'wheel' for up to two or three circles. Afterwards, and assuming that insemination has been successful, the male 'jerks'—'short, sharp, forward and upward movements involving the whole body' (Liley 1966). During a receptive period, and particularly when virgins begin mating, females can accept copulations from several males. The reasons for this, and the consequences for the competing sperm, are examined later in the chapter. Another fascinating comment in Liley's benchmark study of guppy courtship concerns the behaviour he names 'wobble'. He reports that after copulation most females wobble their bodies with large amplitude lateral movements. The body may also be arched slightly. What is most interesting is that this activity is sometimes accompanied by the extrusion of recently inseminated sperm (Liley 1966, p. 42).

Males that encounter a receptive female, particularly in a one-to-one context, only infrequently employ sneaky mating. However, the presence of a rival male causes males to switch some of their effort to gonopodial thrusting (Fig. 4.3). In the wild, or in aquaria where there are unreceptive females present, females can receive up to one sneaky mating attempt per minute (Magurran and Seghers 1994*c*). (Fig. 4.4). Although there are many volumes of papers devoted to guppy courtship behaviour the absolute, and relative, success of sneaky mating behaviour remains shrouded in mystery. It seems curious that an activity that occupies so much of a male guppy's time, and presumably incurs costs in terms of increased predation risk, is still of debatable significance. The literature traces the changing opinions on the matter. At first it was widely assumed that females were unwilling or indifferent consorts that functioned as the receptacle for male reproductive activity. Breder and Coates (1935) wrote that 'actual transfer of material seems only to occur when the male has slipped up to the seemingly unsuspecting female'. However, the development of techniques that allowed researchers to check a female's genital track for recently inseminated sperm soon reversed that view. Clark and Aronson (1951), Kadow (1954), and Baerends *et al.* (1955)

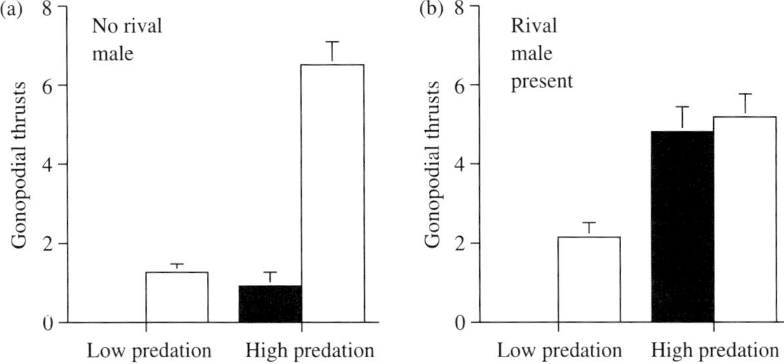

Fig. 4.3 Number of gonopodial thrusts (mean ± s.e.) directed towards either receptive females (open bars) or non-receptive females (solid bars) in (a) the absence and (b) the presence of a rival male. Mean values per 10 min, $n = 32$ males per treatment. Low-predation fish were derived from the Upper Tunapuna River, high-predation fish from the Lower Tacarigua. Most striking is the increase in sneaking towards receptive females by high-predation males under competition. After figure 2.6 in Matthews (1998).

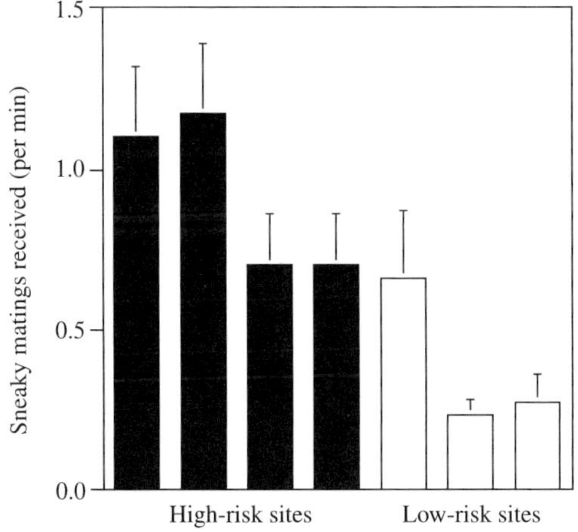

Fig. 4.4 Frequency with which wild female receive gonopodial thrusts from males (per min). Mean values, ±s.e., are shown for seven populations. High-predation (*Crenicichla*) sites are denoted by filled bars, low-predation (*Rivulus*) sites by open ones. After Magurran (1998).

concurred that gonopodial thrusting only rarely resulted in insemination. Liley (1966) supported these observations and noted that the vast majority of thrusts are not followed by jerks and that even when they are, this is no guarantee of successful sperm transfer. This body of research has been pivotal in shifting the perspective from that of a male-dominated mating interaction to a female one. In fact, as it turns out, this was known all along, but had been overlooked in the changing fashions of science. Regan (1913) commented that 'it is of some interest to note that in *Poecilia* and related genera the females welcome the attentions of the males' while Stepanek (1928) was aware of the consensual nature of copulation.

Recent research, particularly that undertaken by Anne Houde and her colleagues, has verified the importance of female choice in guppies, and has done much to elucidate the mechanisms involved. All the evidence suggests that it accounts for most paternity. At the same time researchers have renewed their interest in sneaky mating behaviour, stimulated by the high profile of sexual antagonism and new findings concerning its role in evolution. Kadow (1954) was first to confirm sperm transfer during sneaky mating. Pilastro and Bisazza (1999) mechanically stripped the sperm stores of male guppies at rest and then used these data to estimate the mean number of sperm delivered during a copulation. They found that over 92% of available sperm could be inseminated. Although the number of sperm delivered during a consensual copulation is three times greater than that delivered during a sneak mating there is considerable overlap in the distributions of inseminate size. In 70% of the sneak copulations (which were judged to occur when jerking took place) at least some sperm were transferred to the female (the equivalent figure for consensual matings was 92%). In both cases, the number of sperm delivered correlated with the number of sperm available. Intriguingly, small males delivered a larger proportion of their available sperm during sneaky mating than large males did. Although this study provide unequivocal evidence that sneaky mating can be successful, at least in terms of sperm transfer (see also Pilastro *et al.* 2002), its importance in the wild remained speculative. Iain Matthews and I (Matthews 1998; Matthews and Magurran 2000) made use of the knowledge that female receptivity (and accepted copulations) are confined to the first 4 days of the brood cycle (Liley 1966). We also re-confirmed Liley's observation that sperm can be retrieved from the female's oviduct for no more than 7 days following mating (Kuckuck and Greven 1997). By putting these two bits of information together we realized that the presence of sperm in the genital tract of a female bearing well-developed embryos would be strong evidence that these sperm were the result of sneaky mating. Iain and I dissected the ovaries of 250 females collected from the Tacarigua and Tunapuna Rivers in Trinidad. We used Haynes's (1995) classification of poeciliid embryo development as a gauge of female receptivity. Some 15% of females with well-developed embryos had recoverable sperm, leading us to conclude that these sperm were the result of sneaky mating. Jonathan Evans (Evans 2000) refined the methods of sperm recovery and together with colleagues extended the number of natural populations surveyed. In contrast to the previous study, Evans *et al.* (2003*a*) used a non-lethal method to discern late-term females. This had the advantage that females could be returned to the site of collection. The study showed that 45% of the 376 females, drawn from eight localities, had received sperm through sneaky mating. There was, however, no evidence that the incidence of females inseminated through forced copulation varied systematically across predation regime. This point is revisited in the next section. Houde and Hankes (1997) suggested that one reason why the strong female preference for orange in the Yarra population does not translate into high orange males in the wild is that sneak copulations limit the female's ability to exercise choice. Finally, some recent work testifies to the efficacy of sneaky mating in transferring sperm across as well as within species. Russell *et al.* (2005) collected female poeciliids from two localities where *P. reticulata* and *P. picta* occur sympatrically. Males that have experience of heterospecific females direct many fewer sneaky matings towards them than naïve males do (Haskins and Haskins 1949; Liley 1966; Magurran and Ramnarine 2004). However, their record is not perfect and heterospecific sneaky matings can be observed with low frequency in the wild (personal observation) and in the laboratory (Magurran and Ramnarine 2004). Russell *et al.* (2005) used a genetic method to check for heterospecific sperm in the gonoducts of wild-caught female *P. reticulata* and *P. picta*. Overall, around 4% of females (of both species) had sperm derived from heterospecific copulations. Since neither species of female appears to copulate consensually with the other species (Liley 1966) we concluded that heterospecific sperm were the result of sneaky mating. Incidentally, the same study revealed conspecific sperm in 86% of females but was unable to shed any light on the mode of copulation. The implications of these results for sexual isolation are considered in Chapter 6. The difficulty of deciding, even with the help of finely resolved molecular markers, how sneaky mating translates to paternity, arises because males use both tactics and can rapidly alternate them. Consequently, accurate identification of a father tells us nothing about how the sperm that fertlized his offspring were delivered. It is possible to manipulate encounters between guppies so that some males will only sneak—but this is a simplification of the natural situation. Finally, sneaky sperm must compete with sperm stored from previous consensual matings, as well as with subsequent ones. It will thus require a lot of detective work, paired with careful manipulation and observation, to deduce the significance of sneaky mating in the wild.

What happens to the sperm when they reach the female? The paired ovaries of female poeciliids are fused to form a single organ that fills most of the peritoneal cavity (Constanz 1989). The muscular wall of the ovary, along with its folded inner surface, extend through the short gonoduct (Philippi 1908; Wourms 1981; Constanz 1989). The gonoduct's exit lies just behind the anus and in front of the anal fin. It can be identified by a fleshy papilla (Constanz 1989) though this is more prominent in some individuals than in others (personal observation). Recent research by Kobayashi and Iwamatsu (2002) has shed new light on the process of internal fertilization. Ovarian follicles are connected with the ovarian cavity via a small tract. These micro-tracts become more distinct as the eggs approach the stage where they are ready for fertilization. The diameter of the bottom of this structure, called the sperm storage micropocket (SSP) by Kobayashi and Iwamatsu (2002), correlates with the diameter of the oocyte. As the SSPs enlarge (in tandem with oocyte development), they become populated with more and more sperm. Whether these sperm migrate from elsewhere in the ovary or are inseminated afresh is unclear. By the time the follicle is at its most advanced stage ($\sim$1.8 mm in diameter) the sperm heads are lined up on the bottom surface of the SSP (next to the oocyte), which now consists of a thin layer of epithelial cells. There appears to be no special structure for sperm entry. It seems that the

SSP membrane, follicular layer, and chorion (egg envelope) are penetrated by the sperm at the time of fertilization (Kobayashi and Iwamatsu 2002).

Ever since researchers began investigating the biology of guppies it has been known that females can store sperm. Schmidt (1920) remarked that 'a female whose mate had been removed after copulation might give birth to as many as seven broods at intervals of about a month, the explanation being, that the sperma are stored in the genital ducts of the female and can thus fertilize the ripening ova'. Winge (1922*b*, 1937) confirmed that broods can be produced for up to 8 months in the absence of males. These findings were occasionally overlooked (Purser 1937) and the ability of females to reproduce in the absence of males led to erroneous conclusions about pathenogensis in guppies (Spurway 1953, 1957). The sizes of broods produced solely on the basis of stored sperm decline over time suggesting that sperm are limiting. It is also possible that sperm storage varies in relation to predation regime. A comparison of the duration of sperm storage in two populations revealed that females in the low-predation Tunapuna River had significantly more broods than females in the high-predation Tacarigua River (B. H. Seghers and A. E. Magurran, unpublished study). A number of factors could account for this. Tunapuna guppies have smaller broods, are in a population more prone to female-biased sex ratios, and inhabit a fragmented stream system in which mating partners may be hard to locate.

While the fact of sperm storage is well established, the manner in which stored sperm interact with new sperm remains unclear. Rosenthal (1952) used a mutant strain of guppies to discern how new matings compete with old ones. He discovered that a female who is remated up to 6 days after parturition produces some offspring fathered by the new male. The percentage of replacement, determined by the composition of the brood, declined from 83% from copulations on the first day after brood delivery, to 60% and 36% from those on the 4th and 6th days, respectively. Stored sperm continue to fertilize eggs even in the light of subsequent inseminations. Becher and Magurran (2004) found that 25% of offspring produced by wild-caught guppies over a three month period could be attributed to matings prior to capture. (This experiment did not begin until all females had had ample opportunity to mate with males, from the same population, housed in the same aquaria.) The existence of stored sperm is sometimes viewed as an impediment to research as it can mean that virgin females must be raised *de novo* for each experiment. However, it will be fascinating to finally discover how they compete with fresh inseminates, and whether this is under female or male control.

4.2 Female (and male) choice

Andersson (1994) makes the important point that sexual selection is the process driven by differential mating success among individuals (usually, but not exclusively males). It results in the evolution of secondary sexual characters, such as the peacock's tail, or in the case of the guppy, male coloration. These traits are not the only outcome of sexual selection of course, but they are its most obvious manifestation. The means by which sexual selection is achieved is termed 'mate choice' or 'mating preferences'. (Precopulatory male–male competition can also lead to secondary sexual characters but this appears to be, at most, of only minor significance in guppies). Choice is typically expressed in the form of a behavioural decision whereby a female might opt for the male that has the brightest orange spots or the biggest tail. However, it is becoming increasingly clear that inherent individual differences in the choosing sex, such as variation in the way in which colour is perceived, modulate the decision.

Phenotypic variation in *choosiness* can be dissected into two measurable components (Brooks and Endler 2001*b*): *mean responsiveness*, which is the average responsiveness of a female to all the males she meets (during a receptive phase), and *discrimination*, which is the extent to which she distinguishes among these potential partners. Another way to think of discrimination is as the standard deviation around the mean response. Thus a female who responded to all males in a similar fashion would have low discrimination, which translates into a narrow standard deviation around her responsiveness score. A *preference function* is the descriptor of the relationship between the female response and a given male trait. For example, females may show highest preference for intermediate levels of orange coloration in males (Houde 1987). A useful method of quantifying preference functions is introduced by Brooks and Endler (2001*b*). This uses quadratic coefficients to describe the slope of the relationship between female response and the standardized male trait.

As Brooks and Endler (2001*b*) reveal, females find some male ornaments universally attractive but differ in their preferences for other traits. Their study uncovered strong preferences for large tail area, iridescent area, and the area and chroma of orange spots. In contrast, females varied considerably in the extent to which they found black attractive. Most of the variation among the females could be attributed to differences in responsiveness. A further interesting observation was that although females were consistent in their behaviour, the heritability of mate choice, with the exception of responsiveness, was low and non-significant. Early experience, particularly exposure to particular male phenotypes, is one factor that can influence female mate choice (Breden *et al.* 1995; Rosenqvist and Houde 1997). Mate copying is another (Dugatkin 1992*a*; 1998*a*, *b*; Dugatkin and Godin 1992*c*, 1993, 1998*a*; Brooks 1996; Dugatkin *et al.* 2002). Neither would seem to apply in the Brooks and Endler experiment since naïve virgins were tested.

Virtually all the emphasis on female choice has been directed towards preferences for colour. Olfactory cues have been largely neglected even though research on sticklebacks has revealed that they are used to select partners with complementary MHC profiles (Reusch *et al.* 2001). Recently, Shohet and Watt (2004) discovered that female guppies can distinguish between males on the basis of olfactory cues, but that they rank these males differently when presented only visual cues. As noted earlier the evidence that display behaviour, *per se*, is important is mixed. However, it is possible that the sigmoid display amplifies olfactory cues, or conveys mechanical ones. It is certainly curious that guppies that live in naturally turbid water retain display behaviour in their repertoire. This alternative, or perhaps additional, function deserves investigation. Recent work on MHC loci in *P. reticulata* (van Oosterhout personal communication) supports the contention that these genes play an important, but as yet barely explored, role in mate choice in guppies.

A further topic that awaits resolution is the importance of male size in female choice. Again the evidence is both contradictory and tantalizing. Reynolds and Gross (1992) discovered that larger males had an advantage in no-choice mating trials while Magellan *et al.* (in press) showed that females preferred the larger of two brothers, which resembled one another in colour pattern and differed only in size (size differences were engineered by rearing the fish at different temperatures). Karino and Matsunaga (2002) similarly found that male total length was important. However, an extensive investigation of female preferences in 11 guppy populations uncovered one case where large males were preferred, one case where small males were chosen and nine where there were no preferences (Endler and Houde 1995). Becher and Magurran (2004) discovered that small males were relatively more successful in gaining paternity when fish were allowed to mate freely during a three month period—a result that was replicated in an experiment that used artificial insemination and thus removed all behavioural interactions from reproduction (Evans *et al.* 2003*b*). These studies highlight two issues in the assessment of female choice. First, the manner in which tests are conducted—whether they are choice or no-choice, for example, or whether they present multiple or single cues—as well as the extent to which other variables are controlled, will influence the outcome. This variation in protocol is superimposed on the evolutionary history of the populations concerned. Second, female choice in a choice experiment will not necessarily translate into enhanced paternity for the preferred male once cryptic female preferences, alternative male mating tactics and sperm competition are factored into the equation. This is not to say that female choice is unimportant but rather that it is one of several elements that together determine variation in mating success.

Females may be inconsistent in their preferences for large males. But in fish, where males are concerned, big is often better (see, for example, Coté and Hunte 1989; Pelabon *et al.* 2003). The reasons for this are logical: female fecundity is typically some function of body size (Wootton 1990; Charnov 1993). It has also recently become apparent that bigger—which also usually means older—female fish produce higher quality offspring (Berkeley *et al.* 2004). Haskins and Haskins (1949), Baerends *et al.* (1955), Benz and Leger (1992), and Abrahams (1993) confirmed that more courtship is directed towards larger female guppies. Houde (1997) also notes that large females generally receive more attention, but points out that this is influenced both by sex ratio and by the activity of competing males. A trio of recent papers has unravelled more of the details. Dosen and Mongomerie (2004) found that male guppies associate more with larger females—but do not display more to them. The strength of male preference is correlated with absolute female size, as well as with relative size. Males in Dosen and Montgomerie's experiment were in visual but not olfactory contact with females and therefore had no opportunity to evaluate female reproductive status using pheromonal cues. Their experiment confirms that if males are presented with a binary choice of large versus small, with other variables controlled, big is indeed better. Ojanguren and Magurran (2004) discovered that larger females—both wild-caught and laboratory-reared—receive more sneaky mating attempts when the sexes are able to interact. Although we do not know how successful sneaky mating is, it seems reasonable to assume that larger, more fecund, females offer more tickets in the paternity raffle, as well as providing an easier target for gonopodial thrusting. The relationship between sneaky mating and female size is stronger in guppies from the high-risk Lower Aripo population than from the low-risk Upper Aripo—a result anticipated by field studies that indicate that sneaking is more prevalent in localities with many predators (Luyten and Liley 1985; Magurran and Seghers 1994*c*, see next section). (This does not imply that male choosiness varies between predation regimes—Rodd and Sokolowski 1995). However, in contrast to sneaking behaviour, sigmoid display rate is influenced by female receptivity, with non-pregnant guppies receiving most courtship. Indeed, we found that the frequency of sigmoid displays was unrelated to female size. This result makes sense since only receptive females will engage in consensual courtship. Herdman *et al.* (2004) obtained a similar result in a test that used exclusively virgin females and where the full range of cues (visual, chemical, tactile) was available to males. Males did differ, however, in their approach behaviour, and in the frequency of gonopodial nipping so that, overall, more sexual acts were directed towards larger females. These authors showed that large and small females responded to males in the same way. In the wild, of course, the situation is even more complex since females will receive copulations from several males. Accordingly, it is not just the number of eggs a female can produce, but also the number of competing fathers that will determine a male's chances of siring offspring. Herdman *et al.* (2004) demonstrated that larger females had significantly more sires contributing to their broods. This suggests that it is not always advantageous for males to pursue the largest mate. However, since paternity is distributed unequally among sires, with one male typically fathering a large proportion of offspring (Evans and Magurran 2001; Evans *et al.* 2003*b*), the arithmetic is not straightforward. Furthermore, Smith *et al.* (2002) discovered that males use UV information about females in mate choice, thus confounding the widely held assumption that female appearance is irrelevant. Females also use UV information, but not in the same manner as males (Smith *et al.* 2002).

Another factor that has proved important in male mate choice is familiarity with females. Many animal species exhibit the so-called 'Coolidge' effect, that is a heightened sexual response by males towards novel females (Wilson *et al.* 1963). The term was inspired by an anecdote about a visit by President Calvin Coolidge and his wife to a US Government poultry farm. Mrs Coolidge asked that her husband be informed about the sexual prowess of a particularly fine rooster. The President duly inspected the bird and inquired whether its multiple daily copulations were with different hens. On hearing the affirmative answer the President replied, perhaps apocryphally, 'tell that to Mrs Coolidge!'. A key element of the Coolidge effect is that males need to distinguish females with whom they have previously mated from those they have not. Given that male guppies persistently court females, and attempt repeated sneaky copulations, it is likely that a male will have mated with most if not all the females in his vicinity. This is particularly likely to occur in the small pools that form in streams during the dry season, and in which the same males and females continually interact. In these circumstances familiarity becomes a surrogate of previous mating partner and any unfamiliar female a new mating opportunity (Heinrich and Schröder 1986). In open rivers the situation is different since fish are not constrained, and as males make more switches between shoals than females (Croft *et al.* 2003*a*, *b*, 2004*c*), they have fewer opportunities to become familiar (Griffiths and Magurran 1998) and less need to distinguish old and new mating partners. Kelley *et al.* (1999) confirmed that males will preferentially court females that are unfamiliar, and confirmed that it is the opportunity to become familiar, rather than the population of origin, that is important (Fig. 4.5). Females too show a preference for novel partners ((Eakley and Houde 2004)—see Section 4.4).

A final point of interest is learned preferences by males for particular females. Haskins and Haskins (1949) discovered that naïve male guppies will initially court conspecific and heterospecific females at random. After a few days, however, virtually all attention is directed towards females of the same species. These conclusions were supported by Liley (1966) and by Magurran and Ramnarine (2004) and help illuminate the factors involved in the evolution of reproductive isolation—a topic revisited in Chapter 6.

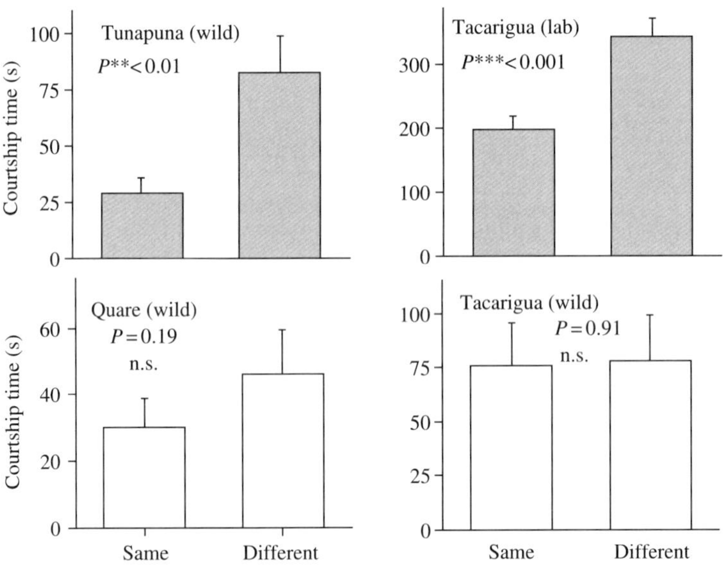

Fig. 4.5 Mean courtship time (+s.e.) by males towards females in the same versus a different shoal (or aquarium). Guppies that associate for 12 days or more have an opportunity to become familiar. Wild Tunapuna fish are isolated in pools during the dry season. This effect was mimicked in Kelley and coworkers' (1999) study by placing Tacarigua fish in aquaria. Males discriminate against familiar females in both cases. In contrast, males make no distinction between females from their own and a different shoal when the guppies are drawn from an open river (Quare and Tacarigua in this example) in which there are ample opportunities to move around and where shoal membership is dynamic.

4.3 Variation in mate choice and reproductive behaviour in relation to risk

In addition to individual differences in female choice within a single population, there are marked differences among populations in the manner in which mates are chosen. As with so much else in the guppy system, predation regime is an important explanatory variable. Finding a mate is something that can usually be delayed, whereas predator avoidance typically is not. For this reason anti-predator behaviour should take precedence over mate choice. These ideas are buttressed by theory demonstrating that degree of choosiness is influenced by the cost of choice (Pomiankowski 1987; Pomiankowski *et al.* 1991). Comparative studies (Breden *et al.* 1987; Stoner and Breden 1988; Houde and Endler 1990; Endler and Houde 1995) confirm that the strength of preference—particularly for orange—is reduced in populations under risk of predation. However, when given a binary choice of a brightly coloured or a drab male, females from high-predation localities still exhibit strong preferences for males with high levels of body coloration (Evans *et al.* 2004*a*).

Differences among populations in choice behaviour are influenced not simply by the female's inherent preferences for different types of males, but also by differences in male behaviour and female perception of risk. Godin and Briggs (1996) discovered that guppies derived from a high-risk locality (the Quaré) reduced their sexual activity and changed their preference (usually from a brightly coloured male to a duller one) when they saw a live *Crenicichla*. In contrast, Paria (low-risk) females did not lessen their interest in males, nor vary their choice of mates. This makes adaptive sense and parallels male risk-sensitive courtship behaviour which is more prominent in populations coexisting with predators (Magurran and Seghers 1990*c*). However, an independent test of the same hypothesis indicates that the outcome is not necessarily as uncomplicated as might be supposed. In contrast to Godin and Briggs, Gong and Gibson (1996) found that females descended from the Paria population reduced their responsiveness (with around 50% becoming sexually unreceptive) and switched from preferring a colourful male to a dull one in the presence of a predatory cichlid. Gong (1997) later showed (in line with Godin and Briggs) that these effects were replicated in guppies descended from a high-risk (*Crenicichla*) population. The reasons behind the different results are unclear but possibly rooted in the rearing regimes to which the fish were exposed, as well as subtle differences in handling and protocol. I also suspect that the way in which the females evaluated the threat was critical. In the Godin and Briggs test, the predator was 37 cm away from the guppies, while Gong's predator was closer and less confined. Higher light levels in the Gong trials would have further elevated the perception of risk (Endler 1987; Reynolds *et al.* 1993). Thus, under high risk (as in the Gong trials) female guppies of all origins (expect perhaps domesticated ones) will become less sexually responsive and moderate their choice behaviour. Indeed, if attack is imminent it is likely that all courtship activity will cease. However, when risk is moderate (as in the Godin and Briggs trials) populations respond differentially, with guppies derived from predator localities needing a lower threshold to trigger response. This hypothesis is readily testable.

There is a parallel with this scenario in the manner in which males adjust their courtship behaviour in relation to risk. It is generally assumed that sneaking is a less dangerous tactic in the face of possible predator attack (Endler 1987). I say assumed because I am unaware of any direct tests, which would in any case be difficult to do, not just for ethical reasons but also because of the considerable challenges of devising a controlled experiment. There is, however, strong circumstantial evidence. First, males from populations where there are many predators generally engage in higher rates of sneaky mating (Farr 1975; Luyten and Liley 1985; Magurran and Seghers 1994*c*; Matthews *et al.* 1997) though this is to some degree confounded by the fact that these same fish typically devote a larger proportion of their time budgets to sexual activities, an outcome that can be attributed to multiple aspects of the ecology of high-predation sites (Magurran and Seghers 1994*c*; Rodd and Sokolowski 1995). Second, high-predation males show greater 'risk-sensitivity' and make a larger adjustment in their behaviour (switching from sigmoid displays to gonopodial thrusting) in the presence of predators (Magurran and Seghers 1990*c*). Finally, Rodd and Sokolowski (1995) demonstrated, in an experiment where fish age and density were controlled, that the gonopodial thrusting of laboratory-reared males was influenced by their population of origin. In line with earlier field studies they showed that males derived from *Crenicichla* localities were more likely to employ sneaky tactics. Sneaking tendency, unlike sigmoid display rate, appears to be heritable.

Females respond to impending predation by increasing vigilance (see Chapter 3) and are generally more risk-averse than males. This suggests that males modify their behaviour not merely to decrease their conspicuousness to predators, but also to exploit the female's preoccupation with predator evasion. Indeed the number of sneaky mating attempts received by females rises significantly when a predator is in the vicinity (Magurran and Nowak 1991). Although predator inspection behaviour is viewed as an activity that fosters cooperation among individuals (Milinski 1987; Dugatkin 1997) this is one context in which the battle of the sexes is fully engaged. Females initiate and lead most inspections; males follow and use the opportunity to attempt sneaky matings (Fig. 4.6). Martin Nowak and I developed a game theoretic model to examine this conflict. We assumed that both sexes have two options—to participate in sexual activity (S) or anti-predator behaviour (A). For simplicity we further assumed that sexual activity means the pursuit of sneaky matings by males—and the avoidance of them by females. P and p are the relative costs for males and females, respectively, of the risk of predation in the absence of anti-predator behaviour. (Upper case letters are used to refer to male payoffs and lower case letters for female ones). M is the benefit (to a male) of sneaky mating; m is its cost for a female. Under extreme risk both sexes should prioritize anti-predator behaviour and abandon all mating activity. This is consistent with observations in the wild and in the laboratory. When risk is less intense, but still severe, the state S,A is the stable outcome. This means that females will avoid predators but not sneaking males, who should opt for gonopodial thrusting. Most interesting is the case of moderate risk, that is when $P < M$ and $p < m$. Here there is no stable outcome but rather a cyclical game in which the sexes switch their behaviour depending on what the other is doing. For example, both might initially focus on predator avoidance (A,A). The male then takes advantage

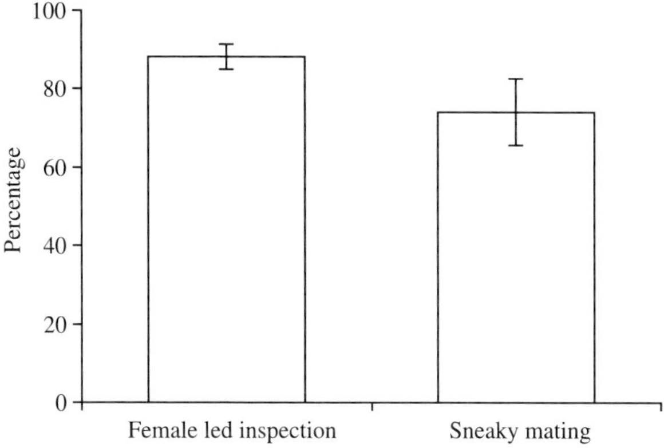

Fig. 4.6 Occasions on which inspections towards a live *Aequidens pulcher* were led by females, and in which sneaky mating took place. Mean percentage (±95% confidence intervals) across 16 trials. Data from Magurran and Nowak (1991).

Table 4.1 Changing payoffs in the battle of the sexes

	Female	
	S	A
Male		
S	-P/ -p	M -P/ -m
A	0/-p	0/0

S denotes sexual activity (sneaky mating by males and the avoidance of it by non-receptive females) and A anti-predator behaviour. The parameters P, p, M, m, describe, respectively, the cost of predation risk for males and females, and the benefit or cost, of sneaky matings to male and females. When both sexes are engaged in mating behaviour—the state (S,S)—then the payoff consists of a greater risk of predation (-P for males and -p for females). If males are sexually active while females avoid the predator—the state (S,A)—then a mating can be forced leading to a payoff M -P for males (who run the risk of predation) and the cost, -m, for females. Alternatively, if males avoid the predator while females avoid males—the state (A,S)—then no mating activity results and zero payoff is obtained by males. Females get -p. Finally, should both sexes employ anti-predator behaviour (A,A) both will receive a zero payoff.

by attempting a mating (S,A). The female's response is to change from predator avoidance to male avoidance (S,S), at which point the male returns to anti-predator behaviour (A,S) and the cycle recommences. This oscillatory pattern is familiar to anyone who has observed the behaviour of guppies in the presence of a threatening but non-attacking predator. It would be interesting to extend the model to encompass situations where females are receptive and where sigmoid displays are employed (Table 4.1).

The notion that a male adjusts his behaviour in response to the female's anti-predator behaviour as well as, or even instead of, the predator has been tested in two studies. Dill *et al.* (1999) found that both male courtship and gonopodial thrusting declined when females were informed about risk (a live *Crenicichla* behind glass). There was no proportional increase in coercive mating. Evans *et al.* (2002*a*), however, demonstrated that the proportion of sneaky mating attempts increased when females had previously been exposed to predation risk in the form of a realistic model *Crenicichla*. Total mating activity (sigmoid displays plus gonopodial thrusts) did not differ between treatments. The different outcomes may be due to the fact that Dill *et al.* used virgin females and a laboratory-reared population of fish, albeit descended from the high-predation Quare population, while Evans *et al.* used wild-caught fish—which were predominately unreceptive females that had had direct experience of predation.

4.4 Multiple mating

There is nearly always a combination of an undiscriminating eagerness in males and a discriminating passivity in females.

(A. J. Bateman 1948)

Multiple mating by male guppies (polygyny) has been reported by many researchers. The oft-observed tendency of males to attempt matings with successive females fits with established notions of promiscuity (Bateman 1948). Indeed, males will even switch to homosexual behaviour in the absence of female mating partners (Field and Waite 2004). However, although it was known right from the outset that females produce offspring fathered by several males (Schmidt 1920), the reasons for this polyandry, and the extent to which it is under male or female control have only recently come under the spotlight.

Haskins *et al.* (1961) used Y-linked colour genes to deduce that wild broods are typically sired by two males. This conclusion is supported by a recent investigation which assigned paternities using microsatellites (Becher and Magurran 2004) (Fig. 4.7). It has also been established that multiple mating is common in the wild (Herdman *et al.* 2004). Kelly *et al.* (1999) conducted a comparative analysis of multiple paternity rates in five high-predation and five low-predation populations. They discovered systematic

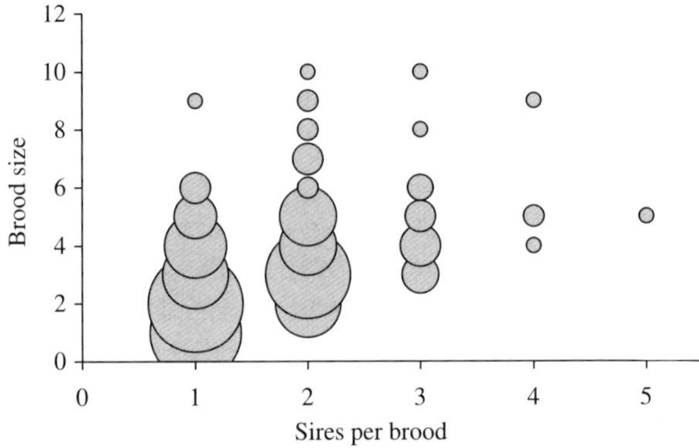

Fig. 4.7 Distribution of brood size in relation to sire number. After Becher and Magurran (2004).

differences with 64 ± 8% (mean ± s.e.) and 25 ± 6% females in high- and low-predation localities, respectively, having broods with more than a single father. Kelly *et al.* (1999) interpret this as a consequence of more intense sperm competition resulting from a predator-mediated shift to sneak copulation (Endler 1987; Magurran and Seghers 1990*c*; Godin 1995) and weakened female preferences (Godin and Briggs 1996). This view is supported by an independent study. Matthews *et al.* (1997) predicted that, as a result of male mating tactics and sperm reserves, levels of multiple paternity should be greater in the high-risk (Lower) Tacarigua than in the low-risk (Upper) Tunapuna. As it turned out females from the high-predation site were indeed more likely to produce multiply fathered broods (70% versus 36% in the Tunapuna). However, as Kelly *et al.* (1999) point out, there are alternative explanations for these results. Environmental conditions can differ markedly between localities; sex ratios also vary—both factors influence mating behaviour (see, for example, Luyten and Liley 1991; Jirotkul 1999*a*). Multiple matings, even unsolicited ones, may result in improved offspring quality (Magurran and Nowak 1991; Evans and Magurran 2000). Alternatively, if the cost of avoiding unwanted matings is too high it may be better to tolerate them (Rowe 1992). Moreover, the fact that the broods produced by females in high-predation sites are larger might lead to higher levels of multiple paternity through a simple sampling effect (Becher and Magurran 2004). Consistent with this is the observation that larger females, which are more fecund, produce broods fathered by more males (Herdman *et al.* 2004). A final complication is that the greater polymorphism of high-predation populations (Shaw *et al.* 1994) makes it easier to detect multiple paternity since genetic markers are typically more variable there (Neff and Pitcher 2002). Fortunately there is now a Bayesian model that can be used to evaluate the power of genetic analyses of multiple paternity (Neff and Pitcher 2002; Neff *et al.* 2002).

The fact of multiple mating by females is incontestable. The reasons for it are only now becoming clear. Jonathan Evans and I conducted an experiment that compared the reproductive output of virgin females that had mated multiply (Evans and Magurran 2000). We found that females that had the opportunity to mate with four (randomly assigned) males, as opposed to a single male (also randomly assigned and presented in the same manner over an identical time-scale) produced more offspring and reduced gestation time. (We subsequently analysed photographs taken of the newly produced broods and discovered that the progeny of multiple matings were also slightly, but significantly, larger (Ojanguren *et al.* 2005)). In addition, the offspring of the multiply mated females had enhanced schooling and predator evasion tactics (Fig. 4.8). Although this study (Evans and Magurran 2000) demonstrates multiple benefits of multiple mating it is not clear to what extent these benefits accrue from the superior genetic material of better males or the female's increased investment in her clutch. The reduced interval between copulation and parturition implies that females are making some contribution. It is also of note that although we recorded matings with all four males on the occasions when multiple partners were offered, on average only 1.62 males contributed to broods. This does not tell us whether differential male success is a product of sperm competition between males or of sperm selection by the female but it does indicate that available sperm are not used equally.

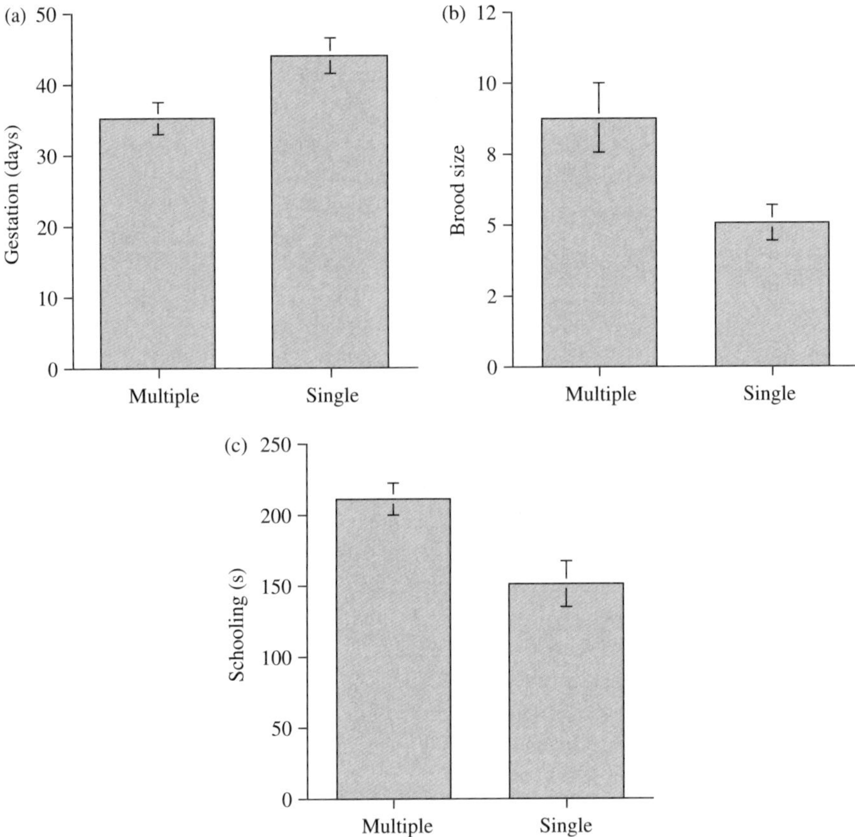

Fig. 4.8 Changes in the gestation length, brood size, and schooling behaviour of newborn offspring sired when females had the opportunity to mate with up to four as opposed to a single male. Mean values and 95% confidence interval shown. Data from Evans and Magurran (2000).

One explanation for the result is that females 'trade-up' (Janetos 1980; Halliday 1983; Gabor and Halliday 1997), that is they mate fairly indiscriminately to begin with, so that some reproductive output is assured, but then re-mate with higher-quality males in an attempt to improve the quality of their brood. This means that a female may delay the development of a brood in 'anticipation' of a further opportunity to mate. Possibly females take advantage of sperm competition to sort out the superior sire but evidence that re-mating increases if the next male encountered is more attractive is provided by Pitcher *et al.* (2003). Trevor Pitcher and his colleagues presented virgin female guppies sequentially with two males of varying ornamentation. Responsiveness to the second male was a function of his attractiveness, where attractiveness was associated with larger areas of orange. The second male to mate gained most paternity and his advantage increased in line with his ornamentation. (Paternity was assigned using a colour marker—see Hughes *et al.* 1999 for details of the method.) Variation in sperm number did not account for these effects though sperm limitation may contribute towards the reduction in the size of broods fathered by a single male.

Angela Eakley and Anne Houde (2004) examined a different (but not incompatible) explanation for multiple mating and asked whether females prefer to re-mate with a novel male. Their experiment supported this hypothesis and further showed that females discriminate against the 'twin'—a brother of similar appearance—of the first male. This result, which is consistent with the work of Farr (1977) and Hughes *et al.* (1999) which showed preferences for rare males, could help explain why wild male guppies show so much diversity in colour pattern. Negative frequency-dependent selection against common phenotypes may help maintain the colour polymorphism that is so characteristic of wild populations (Fig. 4.9).

Sire identity also changes over successive broods. Using microsatellites Becher and Magurran (2004) identified the fathers of offspring produced by female guppies over a 3-month period. Most of these females, which were housed in aquaria with 10 males, gave birth to 3 broods during the investigation. We detected substantial turnover in sire identities between broods. This turnover was, however, less than expected under random mating (Fig. 4.10). Nonetheless, it will reinforce the effect identified by Eakley and Houde and contribute towards the maintenance of diverse colour patterns.

These separate strands of information all imply that females may not be as discerning as Bateman (1948) asserted. Guppies do follow the 'Bateman gradient' of greater variation in the reproductive success of males, than of females (Becher and Magurran 2004). This is due in large part to the constraints that female body size places on reproductive output. But importantly, over a 3-month period—a not unrealistic time frame given mortality rates in the wild (Rodd and Reznick 1997)—the number of mating partners of females matched that of males (Becher and Magurran 2004).

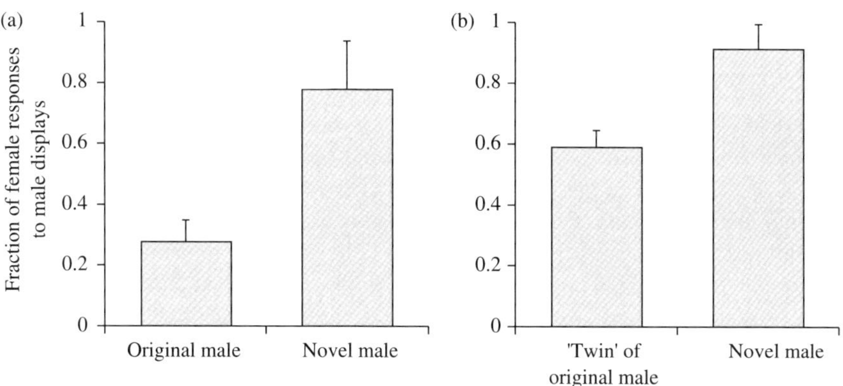

Fig. 4.9 Eakley and Houde (2004) found that (a) females preferred a novel male to a previous mate, and (b) that they also discriminated against the 'twin' of the original male. The fraction of sexual responses was recorded and medians and interquartile ranges are shown.

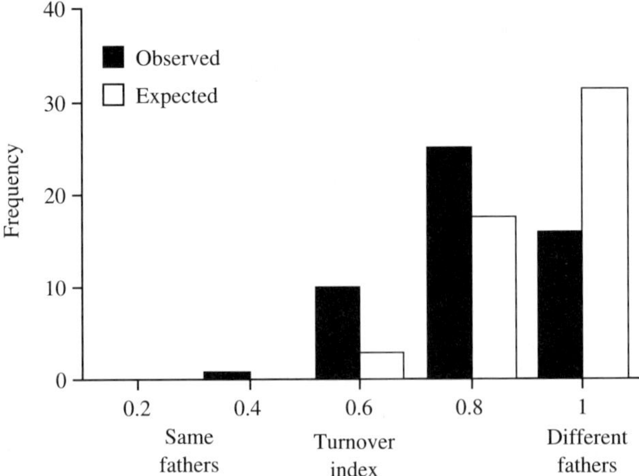

Fig. 4.10 Observed and expected levels of partner turnover during successive broods. Observed turnover expresses the actual number of fathers as a proportion of the total possible number of fathers. For example, the score for a female that has three broods, each fathered by two males, and where five different males were involved, would be 5/6 = 0.83. A value of 1 means that different males fathered each brood. Expected turnover shows the distribution of this turnover index based on 1000 randomizations assuming random mating (from a pool of $n = 10$ males) over three broods with a median of two fathers per brood. The observed and expected distributions of turnover differ significantly (Kolmogorov–Smirnov goodness of fit test, $D_{0.5} = 0.30$, $P < 0.01$) showing that the random mating model can be rejected. Equally, there is a marked bias against the same fathers in successive broods. After Becher and Magurran (2004).

4.5 Sperm competition

These mating patterns provide many opportunities for sperm to compete (Parker 1970). As Birkhead (2000*a*) points out, sperm competition has been known for well over two thousand years. Aristotle, for example, was aware of its existence. The first systematic investigation of sperm competition has been attributed to Robert Nabours (1927), of the Kansas State Agricultural College, who used colour forms to assign paternity in grasshoppers (Birkhead 2000*a*). However, he was pre-dated by Johannes Schmidt (1920), a researcher at the Carlsberg Laboratory in Copenhagen. Schmidt's experiments were carried out in 1917–1918. He first established that female guppies could store sperm by showing that they could produce successive broods without re-mating. He then demonstrated, using Y-linked colour markers, that newly inseminated sperm would out-compete stored sperm:

In explanation of the peculiar state of things here described I have had to recourse to the assumption that the fresh spermatozoa are more agile than the older stock which had been stored for varying lengths of time in the genital ducts of the female, and are thus unable to compete with the former, though they would, if left to themselves, have been capable of fertilising the next batch of ova. The fact that we can, by pairing a female simultaneously with two males of different form, produce, in one and the same brood, offspring belonging to both forms, seems to support this explanation of the selection which takes place among the spermatozoa in the genital ducts

(Schmidt 1920, p. 8)

Other researchers have followed Schmidt in using male colour to deduce the mechanics of sperm competition. His observation that recently inseminated sperm are more competitive has been supported by later work (see, for example, Hildemann and Wagner 1954). The dominance of recent sperm is not restricted to the replenishment of reserves in post-partum females, however. The last consensual mating by a female during a receptive phase also appears to contribute disproportionately to the brood that follows (Evans and Magurran 2001). The distribution of paternity in broods where two fathers are involved is bimodal irrespective of whether sperm transfer occurs in a community tank (S. A. Becher and A. E. Magurran, unpublished data) or during controlled sequential matings (Evans and Magurran 2001). This confirms that the 'fair raffle' model of sperm competition (Parker 1970)—which predicts that paternity will be in proportion to sperm inseminated—does not hold in guppies. Remarkably, this is true even when equal numbers of sperm are contributed through artificial insemination (Evans *et al.* 2003*b*). But it does not tell us whether it is males or females that are engineering this result.

4.6 Cryptic choice?

How the conflict of interest between the sexes is resolved, and its implications for evolutionary rates, is a topic that is attracting a great deal of interest (Birkhead 2000*b*; Birkhead and Pizzari 2002). This chapter has already explored behavioural aspects of sexual antagonism. Chapter 6 will assess its consequences for the evolution of reproductive isolation in the guppy system. The capacity of females to exert cryptic choice during and after copulation is an aspect of sexual conflict that is currently receiving close scrutiny (Birkhead 2000*b*; Birkhead and Pizzari 2002) as well as one that has yielded particularly fascinating results. Recent research has demonstrated that females determine the outcome of mating to a much greater extent than hitherto assumed. Evans and Magurran (2001) discovered that second-male sperm precedence is related to re-mating speed; shorter intervals between matings make it more likely that the second male to mate will dominate the brood. Since mating speed is under female control the implication of this result is that females manipulate ejaculate transfer. This hypothesis was tested by Pilastro *et al.* (2002) in an experiment that related inseminate size to male phenotype. Ejaculate size during solicited copulations was positively correlated with the percentage of orange coloration, showing that preferred males were more successful in transferring sperm. Indeed, the amount of sperm inseminated was unrelated to baseline sperm stores, reinforcing the conclusion that females have an active role in sperm transfer. In coercive matings, by contrast, the relationship between male phenotype and inseminate size broke down. This is additional evidence that females have little control over unsolicited copulations, apart from avoiding them altogether or limiting their duration.

Andrea Pilastro and his colleagues further investigated cryptic female choice in a clever experiment that showed that females can manipulate inseminate size, in favour of preferred males, during copulation (Pilastro *et al.* 2004). Focal males were made to appear either relatively attractive or relatively unattractive by pairing them with a duller or more brightly coloured stimulus male. A virgin female was allowed to observe both males and subsequently allowed to mate with the focal male. On average 68% more sperm were inseminated when the focal male was perceived to be more attractive. As males were unaware of their status, this study provides strong evidence that females bias sperm delivery in favour of preferred males.

Birkhead and Pizarri (2002) predict that cryptic female choice will be more important when mate choice is costly or when reproductive status depends on the genetic compatability of the gametes. Evidence is accumulating that the latter factor is important in guppies. And since mate choice is costlier under predation risk it would be interesting to determine whether post-mating mechanisms are more finely tuned in populations co-occurring with predators. Indeed, as cryptic choice provides a means for females to counteract the inseminates received during sneaky mating by weighting sperm stores in favour of the more preferred males, there are further reasons why it may be relatively more important in high-predation localities. Equally, it would be fascinating to discover whether cryptic choice is correlated with the strength of female choice (Pilastro *et al.* 2004).

The role of the female at the next stage of mating, during fertilization, is less certain. Evans *et al.* (2003*b*) competed sperm directly by artificially inseminating females with equal numbers of sperm bundles from two males. Offspring were genotyped to determine paternity. It transpired that the more ornamented males were relatively more successful. In this population, the Lower Tacarigua, as in others in Trinidad (Houde and Endler 1990; Endler and Houde 1995) preferred males have more orange coloration (Evans *et al.* 2004*a*). In addition, small males had a mating advantage, indicating a possible trade-off between size at maturity and sperm competitive ability. Becher and Magurran (2004) likewise observed that small males, particularly the more orange ones, sired proportionally more progeny. The parallel in the result is particularly striking since the males and females in Becher and Magurran's (2004) experiment were allowed to mate freely. Evans *et al.* argue that sperm choice (by females) is unlikely to explain the mating advantage of smaller males but the reasons why they are favoured in sperm competition remain to be elucidated. Furthermore, irrespective of whether sperm choice or sperm competition is responsible, post-copulatory processes reinforce pre-mating preferences for colourful males.

4.7 Good genes?

The hypothesis that orange coloration is an honest signal of male quality has surfaced repeatedly in the literature. The fact that carotenoid pigments must be gleaned from the diet, rather than being synthesized directly (Goodwin 1984), makes them an important candidate for an honest signal of foraging ability and health (Endler 1980), particularly since they are limited in the wild (Grether *et al.* 1999). Consistent with this is the observation that carotenoid colours are important in male–male competition and in female choice (Kodric-Brown and Brown 1984; Andersson 1994). Males also appear to be under selection to maintain orange spots of a particular hue (Grether *et al.* 2005). Y-linked genes are responsible for some of the variations in male colour traits (Hughes *et al.* 2005). Kodric-Brown (1989) compared the effects of two diets which were identical, save in the levels of caretonoids. Males fed on a diet that included carotenoids had noticeably brighter orange spots than those raised on carotenoid-free food. These brighter males were preferred by females in dichotomous choice tests. Hudon *et al.* (2003) demonstrated that males deposit five to nine times the concentration of carotenoids in their orange spots compared with the rest of the integument.

Grether *et al.* (2004) list three reasons why females should prefer males with more orange. First, there may be a pre-existing bias towards the colour because of the direct health benefits that females gain from being attracted to it. Rodd *et al.* (2002) considered such an origin when they speculated that males might be mimicking orange coloured fruit. This is a plausible explanation for the initiation of the preference for orange—which may have pre-dated the species. However, as there is no correlation between carotenoid availability in the wild and female attraction to orange objects, nor between availability and the strength of female preference for orange (Grether 2000; Grether *et al.* 2004) direct benefits to females are unlikely to account for the continued role of orange in sexual selection. Second, females may avoid mates with sexually transmitted, or other diseases, by shunning males with low orange since carotenoids are redirected to fight infections (Houde 1992; Lozano 1994). Third, more brightly coloured males, being healthy mates, are likely to carry disease-resistance genes that can be passed on to offsprings (Hamilton and Zuk 1982; Folstad and Karter 1992). As yet, the hypothesis that females exploit the relationship between male health and male colour remains unproven. One study, however, provides evidence that it does hold. Grether *et al.* (2004) used the immunological technique of tissue grafting to test the prediction that a higher level of caretonoid intake is associated with an improved immune response. Males fed on a high-caretonoid diet showed a significantly higher rejection rate of the second allograft they were challenged with than males with a low intake of carotenoids. In contrast, there was no improvement in graft rejection in females fed on high levels of carotenoid. The sex difference in response suggests that males pay an immunological cost for their sexual ornamentation. These results show that females who choose males with more intense orange coloration are indeed selecting healthier mates, though it does not prove that these males are genetically superior. The reason why males with more red coloration produce sperm that are advantaged in sperm competition may also have an immunological explanation. Another intriguing point is that the size and distribution of orange spots is under genetic control (Winge 1927; Houde 1992, 1997; Brooks 2000; Brooks and Endler 2001*a*), whereas the chroma (colour saturation) of the spots is affected by diet (Kodric-Brown 1989; Grether 2000). Most studies that attribute a mating advantage to carotenoid colours measure area, rather than chroma. To my knowledge no 'good-genes' mechanism has been advanced to account for female preferences for total spot area, or for the relative area of the body covered with orange spots, though the correlation between area of orange coloration and genetic load (Oosterhout *et al.* 2003*b*) is an intriguing possibility. Finally, guppy males are ornamented by other colours as well as orange. What, if any, indirect benefits these confer on females remains a mystery.

A recent and intriguing study provides support for the idea that females who choose males with more carotenoid markings produce superior offspring. Evans *et al.* (2004*b*) examined the performance of offspring fathered by males that varied in area of orange pigmentation. Artificial insemination was used to ensure that females were unable to assess male attractiveness. Shortly after birth the baby guppies were subjected to a simulated attack from a model avian predator. The time taken to capture pairs of guppies in a hand net was also recorded. Although there was no significant trend among fish in respect of schooling tendency or swimming speed, offspring fathered by males with more orange coloration were most adept at evading capture. This outcome is consistent with a good genes explanation, though it cannot entirely exclude differential maternal investment mediated by ejaculate characteristics.

4.8 Conclusions

The guppy provided some of the earliest insights into reproductive behaviour and the importance of post-mating mechanisms. However, many of the initial puzzles still elude clear answers. For example, although sperm competition in guppies was unequivocally demonstrated almost 90 years ago, the manner in which ejaculates compete with one another is still uncertain. We do not know how stored and freshly inseminated sperm interact, or what favours some inseminates over others. It is also telling that although the guppy has become a model system for investigating questions related to female choice, much remains to be learnt about the proximate and ultimate reasons for that choice. What is clear is that female choices are more complex than a simple reproductive skew perspective might imply. Females mate multiply and switch partners, and gain fitness benefits as a result. But to what extent are these benefits a result of female investment or of male genetic material? Research to date is intriguing but still inconclusive. Females also seem to take molecular and immunological information into account during mate choice but we are only beginning to understand how this might work. Significant uncertainties remain for male behaviour too. Does the sigmoid display have any function other than to exhibit a male's colour patterns? What contribution does sneaky mating make to paternity, and why should males invest so much effort in it? Are male tactics inherited or simply contingent on the social environment in which they find themselves? And why are male guppies, uniquely among poeciliids, so polymorphic? What, exactly, do the different colour patterns signal? Guppies have been instrumental in shaping research on sexual selection and mate choice, but the investigations they have inspired have raised many more questions than answers.

5

Life-history patterns

Guppy populations differ not only in their mating tactics but also in how they make their investment in reproduction. Life-history traits, such as age and size at maturity, number and size of offspring, and life expectancy, vary markedly between populations. Much of this variation is a consequence of predation risk but other environmental factors, such as food availability, fish density, and temperature are also important. Unravelling the reasons for the differences in the way populations and individuals schedule life's events helps us understand how natural selection operates. The investigation of guppy life histories has also been influential in showing that evolution can occur rapidly in the wild.

5.1 Evolution of life-history patterns

It follows directly from Darwin's ideas on evolution that individuals should allocate their reproductive effort in such as way as to maximize the number and quality of offspring they produce. However, the production of progeny means compromise. Individuals must 'decide' when to begin reproducing, how to divide resources between growth and reproduction, and whether to have a few large or many small offspring. The study of life histories is thus the investigation of trade-offs. It is also to a large extent a comparative discipline because the researcher typically wishes to learn how life-history traits, and the trade-offs between them, will be modified when selection pressures change. Guppy populations provide an important opportunity to do this since they avoid the confounding effects that are encountered when comparisons are made among different species. They also offer a near-unique chance to track evolutionary change in the wild.

Although the study of life histories has its origins in Darwin's writing, it is really only in the last half century that the discipline has flourished. David Lack (1944) and Reginald Moreau (1944) made insightful observations on latitudinal variation in the clutch size of birds. In the 1940s and 1950s, however, population biologists were enmeshed in the debate about density-dependent versus density-independent population regulation (Ricklefs and Miller 1999) so the significance of the Lack/Moreau papers was not initially appreciated. Even Caryl Haskins had uncharacteristically little to say about guppy life histories though his work on variation in predation pressure provided the fertile ground that would be later exploited so profitably by David Reznick and his colleagues.

Ricklefs and Miller (1999) place the birth of life-history research, as we recognize it today, in the early 1960s. It seems that a number of complementary influences, including the centenary of *On the origin of species*, and George Williams's groundbreaking reflections on senescence (Williams 1957), prompted investigators to look again at demography and evolution. The year 1966 saw the publication of seminal contributions by Martin Cody, Bill Hamilton, and George Williams that contributed to the new quantitative and theoretical perspective on life-history evolution. A vast outpouring of papers followed. Stearns (1992) and Charnov (1993) provide two overviews of this broad and vigorous topic.

Although life-history evolution has a strong theoretical underpinning, there has been much debate about how best to model it. Most striking in this regard is the notion of r- and K-selection—an early, and influential, approach to understanding the adaptive basis of life histories, though one that later fell into disrepute (Stearns 1992). r-selection and K-selection are metaphors for the contrasting tactics adopted by organisms that are either undergoing rapid population growth or that belong to a population at carrying capacity. The terms themselves are based on the parameters r and K in the well-known logistic growth equation. This r–K approach, which was developed by Robert MacArthur and his colleagues (MacArthur 1962; Pianka 1970, 1974; MacArthur 1972), gained prominence following the publication of *The theory of island biogeography* (MacArthur and Wilson 1967). Reznick *et al.* (2002*a*) credit the r-K paradigm with providing the critical mass needed to establish life-history evolution as a sub-discipline of evolutionary biology.

There are two main reasons why r- and K-selection fell from grace. First, under this scheme species—and/or populations—are typically classified as being subject to one or other form of selection, often with scant regard to their actual population dynamics. In reality most populations fluctuate in size, some dramatically, so a categorical approach is always going to be at variance with the real world. Oversimplification, a charge that can also be made against the predecessor (and related) debate on density dependence and density independence, is a major criticism of r- and K-selection (Stearns 1977, 1992). A second concern is that r–K theory ignores other forms of selection, such as predation and environmental change (Wilbur *et al.* 1970; Reznick *et al.* 2002*a*).

An alternative way to approach life-history evolution is to ask how selection operates on specific age-classes. This 'demographic theory', which was developed by Gadgil and Bossert (1970), Law (1979), Charlesworth (1980) and others, makes predictions about how individual traits will respond given a change in selection. An increase in adult mortality, for example, is expected to result in earlier maturation. The early models were relatively simple and generally ignored density-dependent effects. Later variants became more realistic, inevitably at the cost of increased complexity, and incorporated factors, such as resource availability, indirect effects of predation (Abrams and Rowe 1996), and environmental variability (Kawecki and Stearns 1993). Demographic models, which have dominated investigations of life history in recent years, can be tailored to specific features of an organism's biology and environment. One consequence of this realism, however, is that the simple generalizations of the past no longer hold—the precise way in which density dependence or resource availability is modelled will determine the predictions that are made (Reznick *et al.* 2002*a*). A second consequence is convergence with the approaches inherent in r- and K-selection. Density-dependent selection, as championed by MacArthur, is now evaluated alongside age-specific mortality in contemporary attempts to understand life-history evolution.

The rapprochement between the r–K approach and demographic theory can be clearly seen in investigations of guppy life-history evolution. The initial prediction that predators were driving this evolution by selecting particular size classes of fish, was well supported by empirical data. However, more detailed analyses, which sought to understand the mechanistic basis of this relationship, revealed that the patterns of age-specific mortality were not as straightforward as hitherto assumed. Indeed, it turns out that resource availability, density dependence and competition, as well as the direct and indirect effects of predation are all implicated. Guppy life histories provide a particularly fine example of natural selection in action since they illustrate well how one important cause of mortality (in this case predation) is modulated by other influences. The investigation of guppy life histories started as a test of theory but the data this work generated are, in turn, informing theory. Nature is both more interesting, and more complex, than originally surmised.

The preceding discussion might imply that life-history differences between organisms are fixed. This is not the case. One interesting aspect of life-history evolution is that phenotypic plasticity—the influence of the environment on the organism—may also vary adaptively (Trexler 1989). When individuals from the same strain or population are raised in a range of environmental conditions the 'norm of reaction' (or reaction norm) (Schmalhausen 1949) can be deduced. This reaction norm illustrates how a genotype is expressed under different conditions. Differences between strains or populations in their response to this range of conditions (that is non-parallel reaction norms) reveal a genotype-environment interaction. For example, two populations of guppies might have the same growth rate at a low temperature, but very different growth rates at higher temperatures. Investigations using poeciliids demonstrate considerable plasticity in a range of life-history traits (Trexler 1989). As this chapter will show, guppies are no exception and are characterized by interesting patterns of phenotypic plasticity.

5.2 The role of predators

There is compelling comparative and experimental evidence to show that predation risk drives the evolution of life-history traits in guppies. Reznick and Endler (1982) asked whether variation among guppy populations in life history was consistent with theoretical predictions. It was. They discovered that fish in high-predation sites (that is localities where *Crenicichla* also occurs) mature earlier, and at a smaller size, devote more of their body mass to reproduction (higher reproductive allotment or reproductive effort), produce more, but smaller offspring, and reproduce more frequently than guppies in sites where predation risk is low (the so called *Rivulus* localities). The first indication that life histories change rapidly was provided by the finding that guppies transplanted to a predator-free tributary in the Aripo (Endler's introduction experiment—see Table 3.2) were producing larger offspring and had reduced their reproductive allotment (Reznick and Endler 1982) within 2 years (probably 3–5 generations) of the introduction. A greenhouse selection experiment (see Endler 1980 for details), which ran for 2.5 years, corroborated their results (Reznick and Endler 1982). The key population differences in life-history traits are summarized in Table 5.1.

A large body of work, much of it by David Reznick and his colleagues, has been built on these early findings. The range of populations investigated has increased (Reznick and Bryga 1987; Strauss 1990). Similar patterns of life-history allocation have been recorded in related species (Reznick *et al.* 1992). But it is particularly the transplant experiments, both the initial Aripo introduction by Endler in 1976 (as above), and a later one (1981) in the El Cedro drainage (part of the Guanapo River)

Table 5.1 Population differences in life-history characteristics

Trait	High predation	Low predation	Source
Age at 1st brood (female)	Younger (*87 days Oropouche*)	(*99 days Oropouche*)	Reznick *et al.* 2001*a*
Age at maturity (male)	Younger (*50 days Oropouche*)	(*59 days Oropouche*)	Reznick 1982*b*
Size at 1st brood (female)	Smaller (*168 mg Oropouche*)	(*198 mg Oropouche*)	Reznick *et al.* 2001*a*
Size at maturity (male)	Smaller (*85 mg Oropouche*)	*101 mg Oropouche*)	Reznick 1982*b*
Brood size (per 30 mg female)	Larger (*6.6 El Cedro*)	(*2.8 El Cedro*)	Reznick and Endler 1982
Interbrood interval	Shorter (*26 days Oropouche*)	(*28 days Oropouche*)	Reznick *et al.* 2001*a*
Reproductive allotment	Larger (18% El Cedro)	(11% El Cedro)	Reznick and Endler 1982
Generation time		Slower (*210 days Aripo and El Cedro*)	Reznick *et al.* 1997
Probability of survival for 6 months	Lower (*1–2%*)	(*22–38%*)	Reznick *et al.* 2001*a*
Life expectancy	Longer (*c.1000 days Oropouche*)	(*c.750 days Oropouche*)	Reznick *et al.* 2004
Total offspring production	Greater (*545 Oropouche*)	(*218 Oropouche*)	Reznick *et al.* 2004
Decline in reproductive output	No difference		Reznick *et al.* 2004
Senescence in physiological performance	Rapid decline in fast start response	Slower decline in fast start response	Reznick *et al.* 2004

Notes: Contrasts between high- and low-predation populations are summarized. Illustrative data are included to give an impression of the magnitude of the differences between environments. These data are examples taken from the cited study and not averages, either for that study or for the predation regime as a whole. See individual studies for details, including information on rearing regimes.

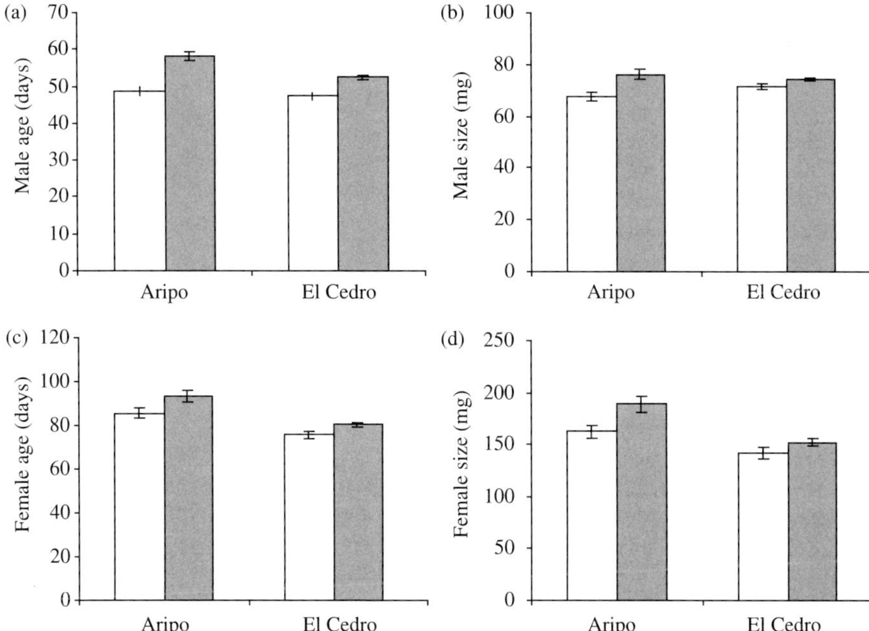

Fig. 5.1 Rapid evolution of life-history traits following release from predation as revealed by two introduction experiments in Trinidad. Data for the Aripo transplant were collected 11 years (*c*.18 generations) after the introduction while the El Cedro transplant was investigated 7.5 years (*c*.13 generations) after the introduction. Life-history traits were measured in fish reared in controlled laboratory conditions. Each graph compares fish in the introduction site (shaded bars) with a control (high-predation) site. Mean values (±s.e.) are shown. The four traits are (a) Male age at maturity in days; (b) Male size at maturity in mg (wet weight); (c) Female age at maturity; (d) Female size at maturity. Data from table 1 in Reznick *et al*. (1997).

by Reznick (see Table 3.2 and Fig. 5.1), that have been influential in revealing how quickly natural selection can operate. In each case there are consistent differences between the descendants of the introduced fish and a control population (representing the ancestral condition) in phenotypic traits—that is, measures made on guppies collected from the wild—such as offspring size, brood size, and body size at maturation (Reznick *et al*. 1990). The change is in the direction predicted by theory. Importantly, the differences between introduction and control persist when fish are reared under controlled conditions in the laboratory thereby demonstrating that evolution has taken place (Reznick 1982*b*; Reznick and Bryga 1987; Reznick *et al*. 1990). One interesting observation is that male traits initially evolve more quickly than female traits, but reach a plateau at around 4 years after the relaxation of predation pressure. This may be because the heritability of male traits is generally higher. The slower evolution in females is probably a consequence of lower genetic variation in size and age at maturity, rather than weaker selection of females. There are also differences among traits, within a sex, in the rate at which they evolve. For example, direct selection acts more strongly on increased age at maturity than increased size at maturity in males.

One of the most striking features of these results is the speed of the documented evolution relative to paleontological levels. Reznick *et al.* (1997) estimate that under the conditions pertaining to the introductions, guppies were evolving at a rate of up to seven orders of magnitude more rapidly than in the fossil record. One reason for this is that directional selection, in this case reduced predation risk, was sustained throughout the study period. The more usual situation in unmanipulated rivers in Trinidad is probably one in which predation intensity rises and falls over time. Most guppies are likely to be subject to an ever-changing selection regime in which no single life-history response is ideal for long. Furthermore, evolutionary rates will decline as the population becomes adapted to the new habitat. The observation of initially rapid evolution in male traits, but only relatively minor changes after 4 years, is evidence for this.

There is no doubt that predators are linked to the evolution of life histories. But how exactly do these instruments of natural selection operate? Demographic theory predicts that reduced adult survival will select for earlier maturation and increased fecundity (Gadgil and Bossert 1970; Law 1979; Charlesworth 1980). A reduction in juvenile survival is expected to have the opposite response. The inference from the transplant experiments and comparative population analyses is that predators are responsible for differences in age-specific survival. This has stimulated a number of attempts to estimate age-specific survival of guppies, both in the wild and in the laboratory. Laboratory tests give conflicting results. Mattingly and Butler (1994) found that *Crenicichla* were less size-selective than previously supposed while Johansson *et al.* (2004) showed that larger guppies were preferentially preyed upon (see also discussion in Chapter 2). A lot will depend, of course, on the relative sizes of the predators, and the composition of the community in which a guppy population happens to find itself. The field tests are also intriguing. Reznick *et al.* (1996*b*) conducted a series of mark-recapture tests in Trinidad which suggested that mortality rates are considerably higher in *Crencichla* than in *Rivulus* localities (approximately 20% mortality per 12 days versus 10% per 12 days in guppies > 18 mm). Juvenile mortality rates are also elevated in *Crenicichla* localities. However, the probability of surviving from birth until maturity in the two predation regimes is nearly identical at 15.2 (*Crenicichla*) and 15.8% (*Rivulus*) because rapid growth and earlier maturation in high-predation habitats help cancel out increased mortality. Mature males in both habitat types suffer higher mortality than equivalent sized females and immature males.

Although this study (Reznick *et al.* 1996b) confirmed, as everyone had suspected, that adult survival is reduced in high-predation sites, it also showed that the mortality *differences* between guppies in *Crenicichla* and *Rivulus* habitats are evenly distributed across all age-classes (Fig. 5.2). According to the predictions of the age-specific mortality hypothesis, larger, older guppies in *Crenicichla* localities should experience much higher relative risk. What might account for this discrepancy between theory and data?

One explanation is rooted in the way models are formulated (Reznick *et al.* 1996*b*). Although most models of age-specific mortality (e.g. Gadgil and Bossert 1970; Law 1979; Michod 1979) are predicated on a change in mortality rates in older individuals,

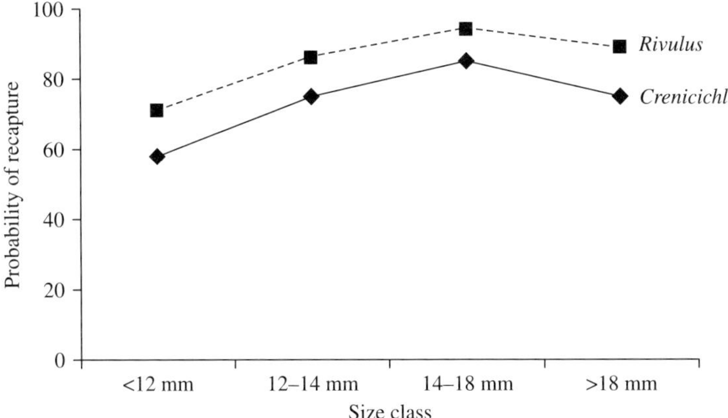

Fig. 5.2 Recapture probabilities of guppies in *Crenicichla* (solid line) and *Rivulus* (broken line) habitats per 12 days. Guppies that coexist with pike cichlids are less likely to survive. However, the curves are approximately parallel showing that mortality differences are constant across all age classes. Mature males are excluded from the analysis. Figure is modified from figure 2 in Reznick *et al.* (1996*b*).

others show that a uniform reduction in survival will result in the evolution of earlier life histories (Kozlowski and Uchmansky 1987). Alternatively, the life history changes might arise as an indirect effect of predation. This could occur if predators reduce density, and thus competition, with the result that per capita resource availability will increase. This is an example of the way in which demographic models have begun to incorporate the density-dependent approach advocated by MacArthur. Finally, there could be inherent differences in the productivity of high- and low-predation sites that, in conjunction with the direct and indirect effects of predation, shape life-history evolution. Discussion of these topics follows. But first I will consider other predators that could play a role in shaping life histories.

As noted in Chapters 2 and 3, predation regimes that guppies experience in the wild are more complex than the usual straightforward *Crenicichla/Rivulus* dichotomies presented in the literature. For example, Endler (1978) classified guppies as belonging to six different predator communities. Little, however, is known about the manner in which these other predators affect life histories. In one exception, the life histories of guppies that occur with the blue acara, *Aequidens pulcher*, a relatively minor predator, were investigated. These fish tend to have life-history traits that are either intermediate of those in *Cencicichla* and *Rivulus* localities, or similar to those in *Rivulus* localities (Reznick and Endler 1982).

The conclusion that predators drive evolution is strengthened by the observation that the same contrasts in guppy life histories between high- and low-predation regimes are evident in different types of predator community. The watershed of the Northern Range in Trinidad marks the boundary between the Antillean and South American zoogeographic regions (see Chapter 2). Rivers that flow north generally lack characins and cichlids (except in cases where these have been introduced). Instead, the lower reaches of the rivers have been colonized by a marine derived fauna. Potential predators here include the gobies *Eleotris pisonis* and *Gobiomorus dormitator*, and the mullet *Agonostomus monticola*. Further upstream these predators drop out, but *Rivulus* is found. Freshwater prawns (*Macrobrachium* spp.) are abundant in northerly draining streams.

This parallel upstream/downstream, low-predation/high-predation contrast provides an opportunity to test for parallelism in life histories. David Reznick and his colleagues (Reznick *et al.* 1996*c*) collected phenotypic information on guppies in six northern drainages. These results were compared with data gathered by Reznick and Endler (1982) in their earlier investigation of south flowing streams. A consistent difference between high- and low-predation localities was recorded for a number of traits including fecundity, offspring size, reproductive allotment, and size of mature males (Fig. 5.3). The magnitude of the difference in some of the traits, such as male

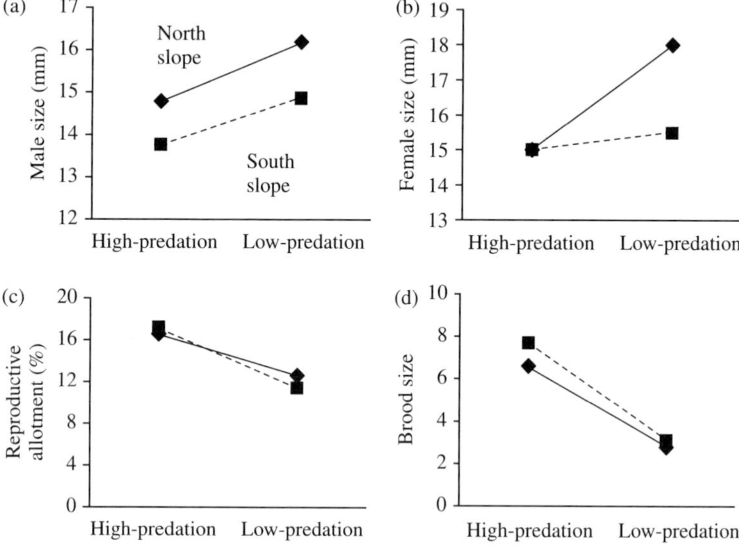

Fig. 5.3 Parallel phenotypes in guppies. This graph illustrates four life-history traits in high-predation and low-predation localities in drainages on the north and south slopes of the Northern Range in Trinidad. North flowing streams are denoted by the broken lines, south flowing streams by the solid lines. *Eleotris* is assumed to be the main predator in northern, high-predation sites, while *Crenicichla* is present in southern, high-predation sites. Weaker predators associated with the low-predation sites are *Rivulus/Macrobrachium* (north slope) and *Rivulus* (south slope). Median values of locality means (taken from table 2 in Reznick and Endler (1982), and table 1 in Reznick *et al.* (1996*c*) (Chapter 4)) are shown. (a) Size of mature males (mm); (b) Minimum size of gravid females (mm); (c) Reproductive allotment of gravid females, that is, the percentage of the dry weight that consists of developing embryos, and (d) Brood size of a medium sized female (i.e. expected fecundity given a female somatic dry weight of 30 mg). The figure follows the style of figure 2 in Reznick *et al.* (1996*c*) but, as medians rather than averages are used, the plotted values differ slightly.

size, varied between north flowing and south flowing streams. This is unsurprising given the manner in which life-history traits vary across seasons and over time (Reznick 1989). One trait, minimum size of reproductive females, did not differ between high- and low-predation sites in the northern streams (Reznick *et al.* 1996*c*). A companion paper (Reznick and Bryga 1996) confirmed that this convergent pattern of life-history evolution has a genetic basis. Furthermore, it uncovered an inherited difference between the high- and low-predation localities in minimum size of reproduction in females. As expected from work on other guppy populations, females derived from high-predation sites in north flowing streams, in this case the Yarra and Madamas Rivers, matured at a smaller size and younger age than their counterparts in low-predation streams. Reznick and Bryga (1996) suggest that the similarity in the female phenotypes of the wild fish (Reznick *et al.* 1996*c*) was due to some unknown environmental effect. Overall, the resemblance between guppy life histories in the two sets of rivers implies that it is severity of the predation threat, and the form it takes, rather than the identities of the predators involved, that underpins selection.

This does not mean, of course, that all predators have an identical influence on life-history evolution. When Chapter 2 assessed various species for their potential as guppy predators *Hoplias malabaricus* emerged as likely to have a significant impact. What is particularly interesting is that it hunts at night. Guppies move to the edge of pools when dusk falls if *Hoplias* is present (Fraser and Gilliam 1992). But it is not just that *Hoplias* will capture and consume guppies. Non-lethal effects on behaviour could also have repercussions for resource exploitation, growth rates, and in turn life-history evolution. In an innovative field experiment Fraser *et al.* (2004) measured nocturnal feeding rates in guppies in the presence and absence of *Hoplias*. The first important finding of this work is that guppies, contrary to popular wisdom, may be active at night and that food consumed then can make a major contribution to their energy budgets. On the other hand, when *Hoplias* is present, feeding virtually ceases after dark because predator avoidance takes precedence over foraging. Fish that are denied the opportunity to feed at night forage more, but court less, during the day (Fig. 5.4). Second, guppies grow at a slower rate if they are prevented from feeding

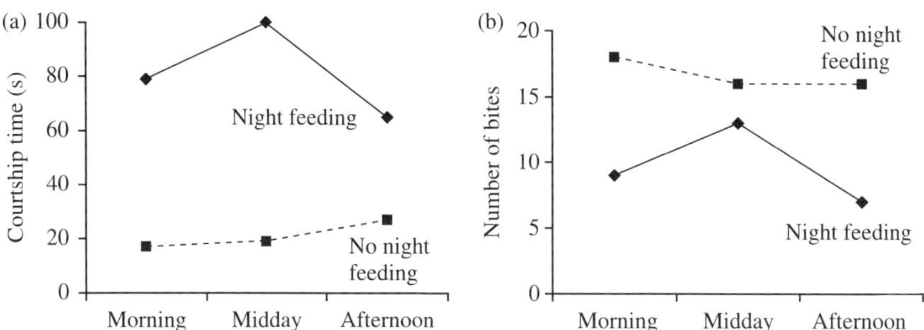

Fig. 5.4 Denial of the opportunity to feed at night leads to (a) decreased courtship and (b) increased foraging during the day. Mean values per 3-min sample are shown. Based on figure 4 in Fraser *et al.* (2004).

at night. If juvenile growth is reduced, theory predicts maturation at smaller sizes (Kozlowski 1992; Abrams and Rowe 1996). This then, is an alternative mechanism that could account for the distinctive patterns of life history found in Trinidadian streams. It is unlikely that any of these mechanisms operate independently. Theory that can integrate the direct and indirect effects of predators, with the effects of resource availability, is eagerly awaited.

5.3 Resource availability and intraspecific competition

It has been known from the outset (Hester 1964) that resource availability affects reproductive allocation in guppies. This observation has at least three implications. First, from a practical perspective, it is essential that feeding rates, fish densities, and so forth are carefully controlled in laboratory assessments of life-history traits. Second, documented differences between wild populations could reflect variation in resource availability, as well as in predation regime. These differences may be phenotypic responses to environmental variation but might also arise as a consequence of environmental selection on life-history traits. Finally, populations could vary in the plasticity of life-history responses.

A reduction in food translates into lower fecundity and reduced reproductive allotment (Reznick 1983), an effect that mirrors the trend seen in the wild during the wet season (Reznick 1989 and see Section 5.5. below). Conversely, brood size increases when food is plentiful. Food-limited females may delay the production of the subsequent broods to allow time to acquire additional resources (Reznick and Yang 1993). Females also adjust the sizes of their offspring to reflect past resource availability. Thus a time of shortage results in the production of fewer, larger babies at less frequent intervals. This could be a form of 'bet-hedging' (Seger and Brockman 1987) against an uncertain environment in which new-born fish need a better start.

Despite the evidence that food availability is linked to phenotypic variation in life-history traits, productivity differences among localities have received relatively little attention. Until recently that is. In the past few years a number of investigators have turned their attention to the environmental characteristics of guppy habitats and begun to ask how guppy life histories might respond to a range of biotic and abiotic variables.

As noted in Chapter 2, Reznick *et al.* (2001*b*) measured the chemical and physical environment, and described macroinvertebrate communities at a number of guppy localities. Guppy life histories were also characterized using standard methods. As before, life-history patterns covaried with predation regime. For example, reproductive allotments in high- and low-predation sites were 16.4 and 13.8%, respectively. There was a tendency for conductivity, nitrates, and turbidity to be higher in the high-predation sites—which were also more likely to be larger streams. Owing the low power of the analysis these differences between high- and low-predation sites were not significant. There were no significant differences in either macroinvertebrate biomass or density though there were fewer smaller invertebrates in low-predation sites. However, gross (periphyton) primary production was significantly higher in high-predation sites at 352 versus 152 C m^{-2} per day.

Indirect interactions occur when one species influences the abundance of another through its interaction with intermediary species (Wootton 1993, 1994). For instance, coral reef fish are more likely to survive predation during the post-settlement period if alternative prey are available at this time (Webster and Almany 2002). Predators, such as *Crenicichla*, eat invertebrates as well as fish, and invertebrates may, in turn, consume guppies. Indirect predator effects could thus conceivably play a role in guppy life-history evolution. Reznick *et al.* (2001*b*) use the absence of a significant trend in macroinvertebrate abundance to dismiss the possibility that the indirect effects mediated through predator/invertebrate/guppy interactions are important. Nonetheless, a detailed investigation of food-web structure in Trinidadian rivers is needed before we can be certain that this is the case.

Indirect effects of predation can be expressed in other ways (Rodd and Reznick 1997). Resources could become limited if local densities of prey increase as a result of their attempts to avoid predation. The concentration of guppies at stream margins in the presence of predators has already been noted. One effect, therefore, of predation risk could be more intense intraspecific competition for limited resources. But it is equally plausible that predators could reduce competition by removing individuals from the prey population. In addition, the partial exclusion of competing species, such as *Rivulus*, from high-predation sites, may increase food availability (Rodd and Reznick 1997; Fraser *et al.* 2004). The smaller body size of guppies in high-predation sites should ease competition further. On balance then, Reznick *et al.* (2001*b*) suggest, food availability will be higher in high-predation sites.

The significance of density effects in regulating guppy populations has recently become clear (Bronikowski *et al.* 2002). Intraspecific competition is the most obvious manifestation of density-dependent regulation though predation and parasitism can also operate in a density-dependent manner. Density dependence, a central feature of the r-K approach, received relatively little attention when the emphasis shifted to demographic models. It was not merely the change in approach that caused density regulation to be neglected, however. Density-independent processes were traditionally assumed (Bronikowski *et al.* 2002) to be of greater significance in the regulation of populations of stream fishes (e.g. Grossman *et al.* 1982; Bayley and Li 1992) although long-term data on species, such as brown trout, *Salmo trutta*, show that density effects can be important, particularly at certain developmental stages (Elliot 1994). Indeed the potential for the density-dependent regulation of fish populations is evident in the structure of stock-recruitment models, such as the well-known Ricker model (Ricker 1954), developed to manage exploited stocks. Experiments on guppies (Barlow 1992) show that population growth, at least in simple environments, is regulated by density-dependent mechanisms including reproductive rate, somatic growth rate, and cannibalism. Bronikowski *et al.* (2002) confirmed that guppy populations in low-predation localities are regulated, in part, by density. Under low densities, adult fecundity increases. High density on the other hand reduces the survival of immature fish.

Productivity differences between high- and low-predation sites contribute a further interaction term to the equation. Grether *et al.* (2001*b*) removed the effect of predation risk by comparing productivity in a range of low-predation habitats that varied in canopy cover. Algal standing crops—measured as Chlorophyll a—differed dramatically between sites. Overall, mean canopy openness explained 78–88% of the variation in mean algal standing crop. Primary productivity, therefore, is reduced in streams where there is good canopy cover. The growth rate of wild females and juveniles was higher in localities that received more sunlight. Males in high light streams were also significantly larger than those in low light ones. 'Common garden' experiments in the laboratory indicated that these differences in growth rate and male size are not genetic in origin. However, females from the high-light localities tended to give birth to larger broods of smaller young—a trait also recorded in downstream (and more productive) high-predation sites. Arendt and Reznick (2005) recently concluded that productivity is more important than predation risk in the evolution of growth rates in female guppies. Resource availability, then, is implicated in the life-history evolution of guppies (Grether *et al.* 2001*b*). The role that competition plays in determining life-history (and other) traits in guppies is also incompletely understood. A combined experimental and modelling analysis of competition in guppy communities, and its interaction with predation, would be illuminating.

5.4 Seasonality

Seasonality also influences the expression of life-history characteristics. During the wet season, which usually runs from May to December in Trinidad, fecundity in Northern Range populations drops and guppies reduce their investment in reproduction. At this time differences in life-history traits between high- and low-predation environments are muted (Reznick 1989; Reznick *et al.* 1990). The reasons for these changes are not entirely clear. Decreases in fecundity and reproductive allotment mirror the changes seen in the laboratory when food availability is reduced (Hester 1964; Reznick 1982*a*). Faster water velocity during periods of high river discharge could displace guppies to the slower river margins. Doug Fraser (personal communication) reports severe scouring effects in upland streams during the wet season; invertebrate populations may be seriously impacted as a result. Silt deposited at these times might additionally make feeding more difficult (Reznick 1989). However, other shifts in life-history characteristics are less easily explained by a reduction in food. For example, whereas there is a reduction in the size of males at maturity, and in the size of females when they first give birth in the laboratory under restricted food (Reznick and Bryga 1987), the reverse is seen in the wild during the wet season (Reznick 1989). Variation in food availability (Reznick 1989), an interaction between nutrition and competition, or a change in predation intensity (cichlid predators in Trinidad tend to breed in the dry season) are just some of the possible explanations for this effect.

Reproductive responses to seasonality are not constant across habitats and species. Alkins-Koo (2000) detected no consistent seasonal trends in female guppy reproductive allotment in the Carlisle-Quarahon drainage in SW Trinidad. Winemiller (1993) found that female reproduction (measured as condition of gonads and proportion of gravid females) in three species of poecilliid in Costa Rica peaked in the early stages of the wet season. Average brood size also increased during the wet season as a result of a marked increase in the fecundity of the largest females. Two of these species, *Poecilia gilli* and *Phallicththys amates*, had poorer diets during the wet season, implying that food resources *per se* are not responsible for the changes. Instead, Winemiller suggests that reproduction is targeted towards periods that are optimal for juvenile growth and survival. Given the complex seasonal variation in the hydrology and ecology of neotropical poeciliid streams (Chapman and Kramer 1991*a*, *b*; Winemiller 1993) it is unlikely that a single environmental variable can explain changes in reproductive effort.

5.5 Plasticity and the social environment

Phenotypic plasticity in the life-history traits may also be a response to the social environment (Rodd *et al.* 1997). Rodd *et al.* (1997) raised juvenile guppies, derived either from a '*Crenicichla*' or a '*Rivulus*' stock, in two demographic settings. The first of these, two adult females plus two adult males plus two juveniles, was designed to mimic the situation in a high-predation environment. The other—five adult females plus two adult males plus two juveniles—resembled the population structure in a low-predation site. (The same approach was used by Rodd and Sokolowski 1995 to examine the consequences of rearing environment for sexual behaviour). Helen Rodd and her co-workers detected an interaction between ancestry and environment consistent with evolved differences in plasticity. For example, female guppies descended from the '*Rivulus*' stock adjusted traits, such as number and size of offspring, in response to a change in social environment, whereas those derived from a '*Crenicichla*' stock did not. There was also more flexibility in the responses of males from a '*Rivulus*' source. Rodd *et al.* (1997) interpret these results in the light of individual reproductive success.

5.6 Plasticity and predators

Predator cues have been shown to influence the behaviour and morphology of many invertebrate and vertebrate species. As Chapter 3 noted, fish learn to respond to novel predator cues if these are paired with stimuli, such as fish alarm substance that are already recognized as dangerous (Magurran 1989; Suboski *et al.* 1990). Chemical cues from predators can also trigger inducible morphological defences. Crucian carp, *Carassius carassius*, a species that is distributed widely across Europe and Asia, and a close relative of the goldfish, develops a deeper body in the presence of its natural predator, the northern pike, *Esox lucius* (Brönmark and Miner 1992; Brönmark and Pettersson 1994; Brönmark and Hansson 2000). A deeper body makes a carp more difficult for pike to capture and handle, and if they have a choice, these predators will direct their attention towards slender prey. There is a cost to this defence too—carp with deeper bodies experience higher drag when swimming (Pettersson and Brönmark 1999). But since the defence is inducible this cost is borne by only those fish that occur with pike.

Life-history traits can also be shaped by predator cues. This effect is particularly well documented in invertebrates. *Daphnia* spp. for instance, will opt for early or late maturation depending on the identity of the predator leaving chemical cues in the vicinity (Stibor 1992; Weider and Pijanowska 1993; Reede 1995; Sakwinska 1998). There have been fewer investigations of predator induced reproductive plasticity in vertebrates but in one example the toad *Bufo americanus* was found to metamorphose at a smaller size in the presence of an odonate predator (Skelly and Werner 1990). The possibility that guppies might also adjust their life histories in response to cues from predators has recently been assessed (Dzikowski *et al.* 2004). Females were exposed to chemical cues, visual cues, chemical plus visual cues, and chemical plus visual and tactile cues from an African cichlid species (*Aulonocara nyassae*) for 18 days. Tactile cues occurred when the cichlid attempted to attack the guppies through the mesh wall of their holding tank. A further control group of females received no cues at all. All females exposed to predator cues produced significantly more offspring in the first brood than the control females. There was no difference between cue types in this response. However, the differences between the exposed and control females were no longer evident by the time the second brood was born. Since the females were virgin at the outset of the experiment, and first mated at around the time they were exposed to the predator cues, the result probably reflects the same underlying mechanisms that are brought into play when multiple mating occurs (Evans and Magurran 2000 and see Chapter 4). It is therefore interesting that there were no differences among treatments and control in gestation period.

Dzikowski *et al.* (2004) used a domesticated strain of guppies (the 'red cobra' variety) and an alien predator. As noted several times elsewhere in this book, 'pet shop' guppies generally have much weaker anti-predator responses than fish derived from wild populations. Furthermore, reactions tend to be strongest towards co-evolved rather than alien predators (Chapters 2 and 3). The possibility that there might be population specific plasticity in reproduction as a result of exposure to predator cues is tantalizing. It has previously been shown that guppies adjust their brood sizes in response to changing fish densities (Warren 1973*a*, *b*; Nishibori and Kawata 1993). Experiments to disentangle direct and indirect predator effects in inducing modification of life-history traits would be both straightforward and rewarding.

5.7 Temperature effects

Despite the current emphasis on climate change—and its negative implications for biodiversity (Thomas *et al.* 2004)—there has been little consideration of the way in which thermal regime might influence guppy life histories. In part this deficiency is because there is relatively little variation in the water temperature of the Northern Range streams that guppy researchers focus on (see, for example, table 1 in Grether *et al.* 2001*b*).

Researchers also usually control temperature carefully when running experiments in the labotarory. But there can be striking diurnal and seasonal differences in the temperature of guppy habitats in Trinidad. Alkins-Koo (2000) reports maximum diurnal temperature ranges at her study site in SW Trinidad of 10 °C for air temperatures and 7.5 °C for surface water temperatures. The seasonal range of water temperatures is 20–28 °C. Kenny (1995) comments that 30–32 °C is not uncommon in standing, exposed water. Guppies, even within the Northern Range, may be trapped in small pools during the dry season and thus experience high temperatures for at least part of the year.

Guppies can be acclimatized to a broad range of conditions in the laboratory and temperatures only become lethal once they exceed 40 °C (Chung 2001). One place in Trinidad where guppies might be expected to experience severe physiological stress on a regular basis is Pitch Lake at La Brea in the SW peninsula. Pitch Lake is a natural asphalt deposit. Notwithstanding its passing resemblance to a large car park—albeit one that would soon devour any vehicle unwisely left there—Pitch Lake is one of Trinidad's main tourist attractions. Pitch Lake was 'discovered' in 1595 by Sir Walter Raleigh who used its asphalt to caulk his ships. As the surface of Pitch Lake is uneven water accumulates in hollows. The resulting pools support simple communities consisting of guppies, *Rivulus* and the predatory leaf fish, *Polycentrus schomburgkii*. What is particularly remarkable is that the surface temperature of these pools can reach 41 °C (Kenny 1995). Given the shallow water it is also probable that water temperatures drop dramatically in the evenings with the result that there is high variation as well as high absolute temperature. As far as I am aware no one has investigated the life histories (or behaviour) of Pitch Lake guppies. Of interest is the observation that adult sex ratios are female biased in Pitch Lake (Pettersson *et al.* 2004)—a result that might be due to the fact that male guppies are particularly affected by high temperatures (Gibson 1954). The juvenile sex ratio at Pitch Lake is not significantly different from 50:50 (Pettersson *et al.* 2004).

The possibility that temperature might be implicated in guppy life-history evolution was raised by Robin Liley and Ben Seghers (1975). Liley and Seghers documented a significant negative correlation between stream temperature and body length ($n = 19$ sites: males: $r_s = -0.5$, $P < 0.05$; females $r_s = -0.66$, $P < 0.01$). They then asked whether these differences persisted under standard conditions. Fish were drawn from the Upper Aripo—an upstream, low-predation, but lower temperature locality, and from the Guayamare—a lowland, high-predation, higher temperature locality. Guppies were raised at two temperatures 23 and 28 °C; all other features of the rearing environment were kept constant. An interesting interaction between origin and thermal regime emerged. Both sexes of both populations grew larger under the low temperature conditions. Upper Aripo males were consistently larger than the ones from Guayamare, particularly at higher temperatures. Guayamare females in contrast were larger than males at both temperatures. These results led Liley and Seghers to conclude that observed population differences in fish size derive both from genetic differences and from a phenotypic response to environmental temperature. Since lowland rivers tend to have little canopy cover, and are therefore more productive as well as warmer, resource availability is a further factor in this relationship.

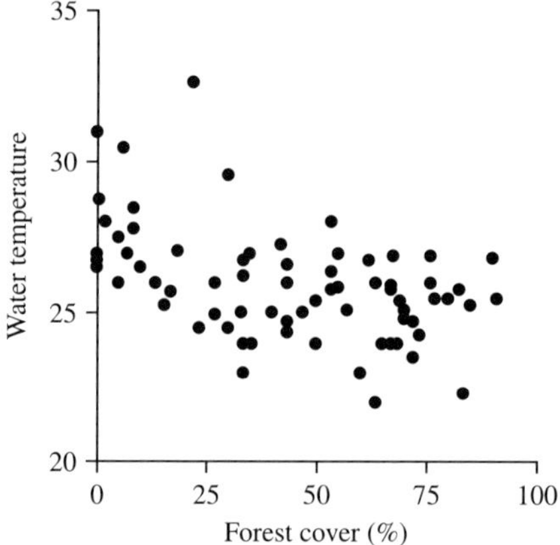

Fig. 5.5 Water temperature (°C) is negatively correlated with forest cover ($r = -0.48$) in Trinidadian guppy streams. Based on figure 1 in Magurran (2001).

Lars Pettersson and I (unpublished data) used a split brood design and measured a greater array of life-history variables in a comparison of two populations (Upper and Lower Aripo) raised in two temperature environments. We found that the populations responded differently to the thermal regimes—a result suggestive of differences in reaction norm. There is thus considerable potential for investigating thermal aspects of life-history evolution in guppies. Parallel research on mosquitofish (Meffe 1991, 1992; Meffe *et al.* 1995) is further confirmation that it will be rewarding.

A final twist to the temperature/life history story is provided by the finding that temperature influences behaviour in guppies (Weetman *et al.* 1998, 1999). What is particularly interesting here is the indication that guppies respond more vigorously to the threat of predation at higher temperatures. The temperatures used in these experiments—22 and 26 °C—are well within the range observed in the Northern Range in Trinidad. But it is not simply that predator defence becomes more urgent at higher temperatures—food requirements are also greater then. A guppy living in a warmer locality might have fewer opportunities to spend time foraging to meet its heightened energy demands if it is also under constant predation risk. It is thus conceivable that the greater productivity of the downstream, high-predation (and warmer) sites (see Fig. 5.5) is not freely available to guppies. These temperature related effects could reinforce the foraging reductions observed by Fraser *et al.* (2004 and see above) in response to the nocturnal habits of *Hoplias*. All speculation of course, but something that might be worth investigating.

5.8 Aging and senescence

Alex Comfort, best remembered for his book *The joy of sex*, also did pioneering work on senescence using guppies (Comfort 1960, 1961, 1963). The advantages of the guppy system that attracted Comfort have recently been exploited by Reznick and his colleagues (Reznick 1997, 2004; Reznick *et al.* 2001*a*, 2002*b*, 2004; Bryant and Reznick 2004) in a series of innovative analyses of aging and senescence in vertebrates. This work shows that the manner in which animals grow older, and the pattern of life expectancy, is not as straightforward as classical theory predicts.

Bryant and Reznick (2004, p. 55) follow Abrams (1993) in defining senescence as the 'decrease in fitness with age caused by physiological degradation and manifested either as reduced age-specific survival or age-specific fecundity'. To put it another way, older individuals are more likely to die and less likely to reproduce. There are two types of mortality associated with senescence. Extrinsic mortality, due to external events, such as predation and disease, influences the evolution of life-history patterns, including the aging process. Intrinsic mortality is the inherent difference between populations or species, in the rate at which individuals age. These differences will arise, at least in part, through the selection exerted by extrinsic mortality. In practice it can be difficult to cleanly separate the two forms of mortality. For example, less proficient anti-predator behaviour in older individuals might reflect an inherent decline in escape responses as well as the foraging preferences of predators.

Bryant and Reznick (2004) used a mark-recapture protocol to estimate intrinsic mortality in two wild populations of guppies. One of these, a tributary of the Quare, is a historically low-predation population. El Cedro, the other, is the result of an introduction experiment (see Table 3.2) in which guppies descended from a high-predation stock were transplanted to a low-risk site. *Rivulus* is the only other species of fish found in both places. Patterns of senescence in high- and low-predation environments cannot be compared directly as the two types of mortality are confounded. Moreover, there are too few older fish in high-predation habitats to permit a quantitative analysis of senescence. (Guppies are 20–30 times less likely to survive for 6 months here than in low-predation habitats (Reznick *et al.* 1996*b*, 2001*a*)). Bryant and Reznick's comparison therefore controls for, though not entirely removes (see below), extrinsic mortality as a result of predation, and provides a comparison in which fish can survive long enough for inherent differences in senescence to be expressed. More than 4500 fish of various ages were marked and sites were resampled six times. All sampled fish were returned to their site of capture.

Their results uncover a legacy of the ancestral history of the guppies. Females in the El Cedro population descended from a high-predation stock, experience a significant increase in mortality rate at 6 months of age—a time when they will be producing only their second or third brood. Quare females, by comparison, do not suffer increased mortality until 16 months. The pattern in males is rather different. Males have higher mortality rates than females. This could be due to their continuing vulnerability to *Rivulus* (Liley and Seghers 1975; Mattingly and Butler 1994), since, unlike females, males cannot escape from these gape-limited predators by growing large. One result of this sex difference is that females are twice as likely to survive over an 8-month period. Males descended from 'high-predation' founders also have higher mortality rates than males in the naturally low-risk population. However, in contrast to the female case, there is no population difference in the pattern of senescence. This might be due to the fact that absence of males in older age classes erodes the power of the analysis to detect an inherent difference. Alternatively, the sexes may genuinely differ in the manner in which their aging responses are shaped by predation risk. One intriguing dissimilarity between the sites is the high prevalence of an unidentified infection in El Cedro guppies. Bryant and Reznick speculate that the immune system might be involved in the trade-off against the early life history associated with high-predation localities. This could mean that guppies that evolve towards a more 'r-selected' lifestyle in response to an increase in predation risk will be less able to fight infections in later life. The issue is further compounded by the observation that the diversity of pathogens correlates with the diversity of fish, that is, there are more disease organisms, as well as more predators, in the lower stretches of rivers.

One drawback of comparative analyses of wild populations, particularly where only a few sites are involved, is that observed differences in the trait of interest may be confounded by all the other features that distinguish the localities. The solution is to dovetail field observation with carefully controlled laboratory experiments.

Reznick *et al.* (2001*a*, 2004) undertook a series of experiments in which fish derived from a high-predation and a low-predation locality in each of the Yarra and Oropouche drainages (thus yielding independent pairs of sites) were raised under high and low food availability, thereby mimicking the variation in resources that would be available in the wild. Reznick *et al.* (2001*a*) observed a decline in the size of individual offspring produced by older females. Interbrood interval also increased with female age, and in fish that had produced four or more litters could be as much as 50 days. Interestingly, reproduction seems more irregular in low-predation than in high-predation fish. The majority of females tested died shortly after the production of their final brood suggesting that a long post-reproductive existence is not the norm. The combination of an earlier start to reproduction, a shorter interval between broods, and larger brood size ensures that, all other things being equal, fish from high-predation populations are more fecund (Fig. 5.6). Indeed, under optimal conditions, a high-predation female can give birth to more than 1000 offspring (Reznick *et al.* 2004) (Fig. 5.6). Brood size initially increases in line with body size and then levels off about 7 months after maturity as asymptotic body size is achieved, before declining in older females.

Varying food ration is one way of controlling for ecological differences between sites that also differ in predation risk. Populations that experience lower predation also generally have lower food availability. A reduction in food tends to increase reproductive lifespan, but reduce fecundity (Fig. 5.6). However, variation in food alone cannot explain the observed differences in senescence.

There are two bodies of theory that make predictions about the evolution of senescence in relation to differences in extrinsic mortality rates. Peter Medawar's (1952) *mutation accumulation* hypothesis suggests that senescence is the by-product of selection to remove deleterious mutations. Any mutations that are expressed early in life will be rapidly weeded out of the gene pool where-as late-acting mutations, which are relatively protected against selection, will accumulate with detrimental

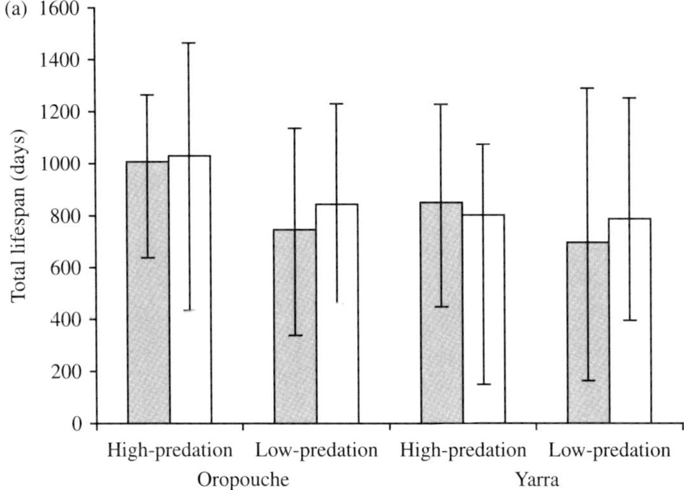

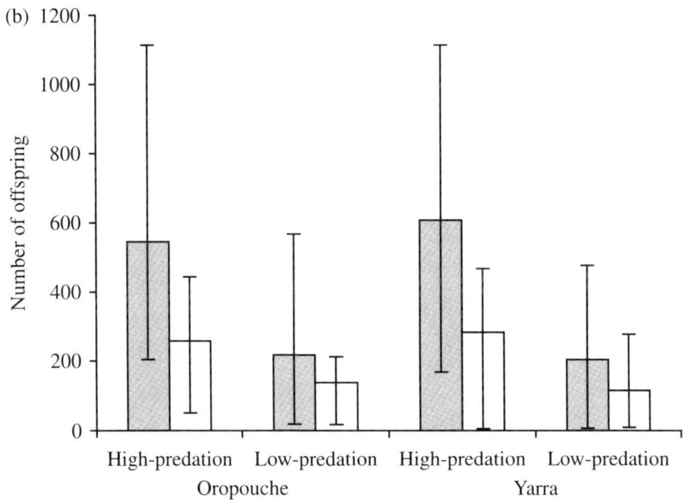

Fig. 5.6 Total lifespan and offspring production by guppies in relation to origin and food availability. Reznick *et al.* (2004) raised second generation (laboratory-bred) guppies derived from paired high- and low-predation localities in two independent drainages (Oropouche and Yarra) under two levels of food availability. Guppies in high food treatments (shaded bars) received food in line with its availability in more productive habitats (usually downstream, high-predation sites) while fish in low food treatments (open bars) had a more restricted diet reflecting food availability in less productive habitats (usually upstream, low-predation sites) in Trinidad. This design allowed the researchers to separate resource level and historical predation risk. (a). Mean total lifespan in days. Error bars represent the observed range. (b). Mean (with range) number of offspring produced by a female over the course of her life.

consequences for older individuals. George Williams (1957) predicted that increased mortality will select for earlier maturity and more investment in reproduction by younger individuals meaning that, as a result of *antagonistic pleiotropy*, there is reduced investment in maintenance in later life. It follows, in both cases, that senescence should be accelerated in populations, or species, with higher extrinsic mortality (Reznick *et al.* 2001*a*, 2004). The early reproductive decline in female El Cedro guppies (Bryant and Reznick 2004) seems to be consistent with this.

Senescence is one of those terms with a deceptively simple meaning. In fact, there are many different ways in which it is possible to become old (Abrams 2004). Reznick *et al.* (2004) asked how age related changes are manifested in three different variables—cessation of reproduction, reproductive output, and physiological performance—that together influence fitness. Their results show that the relationship between the onset of senescence and extrinsic mortality (predation status) of a population is complex.

Reznick and co-workers' (2004) high-predation females not only began to reproduce at an earlier age, but also continued to produce offspring for longer than low-predation females. This is contrary to the expectation that fish that experience low rates of extrinsic mortality in nature will have an extended reproductive lifespan. A second finding was that high-predation females produced higher numbers of offspring throughout their lives. Again, there is no support for the prediction that the rate of decline in reproductive output will be lower in females that evolved in less risky habitats. On the other hand, physiological senescence, in the shape of the fast start escape response, was more evident in females from the predator-rich sites. Although high-predation females had much faster escape responses when young (*c.*1 year of age), there was no difference in response between population pairs in older fish (>2 years).

How might this discrepancy between theory and data be accounted for? One possibility is that density effects could mediate the outcome (Charlesworth 1980; Abrams 1993). For example, high mortality by predators might free up resources that are especially beneficial to older age classes. As Reznick *et al.* (2004) observe, not enough is yet known about density regulation in guppies to test this theory. A significant interaction between food availability and predation status in one of the replicates (Yarra) nonetheless implies that resource/density factors could be important. Other explanations include the possibility that the sharp increase in fecundity after maturity, particularly in high-predation fish, counteracts some of the differences in mortality rate and thereby dampens differences in senescence (Williams 1957). Reznick *et al*'s (2004) study illustrates well, how carefully executed empirical studies that yield unexpected results can challenge existing theoretical preconceptions. The stage is set for reciprocal advances in theory and experimental test in this exciting field.

5.9 Conclusions

Our knowledge of life-history evolution in guppies has increased spectacularly over a relatively short period. The pivotal role of predators as agents of natural selection has been established beyond doubt. Field transplant experiments have demonstrated not only that evolution occurs, but also that it is remarkably swift. Analyses replicated across drainages and predator communities, as well as in the field and laboratory, confirm that the findings are consistent and repeatable. As predicted by theory, guppies that coexist with fierce predators, such as the pike cichlid *Crenicichla*, mature at an earlier age and devote more effort to reproduction.

And yet, this appealing simplicity masks an undercurrent of confounding variables and confusing results. Although experimenters strive to match sites as closely as possible, low- and high-predation localities often differ in productivity, temperature, and in levels of both intra- and inter-specific competition. There are subtle indirect effects of predation. Moreover, the straightforward predictions of demographic models are not as neatly fulfilled as was previously thought. Documented patterns of aging are also at variance with some of the classical predictions of senescence models.

Density-dependent population regulation has recently been revived as a plausible explanation for life-history evolution. Models of age-specific mortality are being adapted to incorporate density effects. These issues are challenging to study as density dependence may be more important at some life stages than others, or only evident at extreme population sizes. It is also unclear whether density dependence is a universal feature of guppy populations. Bronikowski *et al.* (2002) for instance, argue that density regulation does not operate in high-predation environments. Teasing apart the intertwining influences of predators and productivity remains a goal for the future. Competition and other correlates of density, such as disease, also deserve much greater attention.

6

Evolution of reproductive isolation

Guppy populations evolve rapidly following a change in predation risk. The rate of evolution is high, particularly when compared with levels in the fossil record (Reznick *et al.* 1997, chapter 5). Marked genetic divergence between fish in the eastern and western (Oropouche and Caroni) drainages is indicative of a long period of separation, and means that there should have been ample opportunity for isolating mechanisms to arise in allopatry. And yet, it appears that isolation is at best weak and that guppies from different populations will interbreed freely. This interesting puzzle has begun to attract more attention (e.g. Endler 1995; Magurran 1998; Brooks 2002) and has stimulated several recent investigations. In this chapter, I review the evidence for reproductive isolation at each stage of the reproductive sequence, that is, before, during, and after mating. One of the central questions in speciation biology is the order in which isolating mechanisms emerge (Coyne and Orr 2004). Too often the issue is clouded by the fact that these mechanisms are evaluated in 'good' species, with well-established boundaries and where several isolating barriers are already in place. As guppy populations are at a relatively early stage of diversification they provide an opportunity to examine the potential of a range of pre- and post-mating mechanisms (Mayr 1942, 1963). It is of course one thing to identify isolating barriers that operate at the initial stages of divergence and another to establish which barriers persist throughout the speciation process (Coyne and Orr 2004). Here data from other poeciliids can be illuminating. Investigations of guppies also show how the conflict of interest between the sexes influences reproductive isolation. But I begin by revisiting the assumption that guppy populations in Trinidad have not yet set out on the road to speciation.

6.1 The case against incipient speciation

Guppies from different populations mate readily and produce viable offspring
(Houde 1997, p. 151)

. . . there is no evidence that these (Trinidadian guppy) populations are reproductively isolated
(Magurran 1998, p. 278)

. . . there is no sign of genetical (postmating) incompatibility among populations
(Endler 1995, p. 28)

My own experience, in common with that of other researchers, is that when guppies from different Trinidadian populations are given the opportunity to breed, they will do so without any obvious impediments. For example, there is no reduction in fecundity relative to within population crosses, when females from the Caroni drainage are mated with males from the Oropouche drainage—and vice versa (Russell 2004). Males seem to court females without regard to their origin (see Section 6.3). And although females have a slight—but significant—preference for native males (Houde 1988*b*; Endler and Houde 1995) they do not appear to discriminate against partners from a genetically divergent stock (Magurran 1998). For example, Magurran *et al.* (1996) examined female preferences in guppies from the Tacarigua and Oropouche Rivers. These fish are as likely to solicit matings from males belonging to the alien drainage as from their own population (Fig. 6.1).

Guppies above and below barrier waterfalls show consistent differences in a range of adaptive traits. Fish that move downstream from a safer to a more dangerous locality will be vulnerable to predation and thus have fewer opportunities to breed. Any hybrid offspring that are produced may have inferior behaviour and inappropriate life histories. Guppies, particularly males, that make the reverse journey upstream could be at a disadvantage in sexual selection.[2] 'Ecological speciation' (Schluter 2000), in which reproductive isolation emerges as a by-product of ecological differences, is therefore a plausible process for guppy populations experiencing different predation regimes. However, reproductive isolation is unlikely to evolve between contiguous low- and high-predation populations for at least two reasons. First, within-stream environmental gradients may be relatively short-lived and there may be insufficient time for isolating mechanisms to become established. Endler (1995) suggests that predation regime boundaries in Trinidadian streams could persist for fewer than 1000 years. Waterfalls and riffles can be modified by flooding, siltation, and landslips, and ecological communities will change if predators colonize further upstream or become locally extinct.

The second factor that impedes ecological speciation is gene flow (Endler 1995). Colour markers (Haskins *et al.* 1961), allozymes (Shaw *et al.* 1991, 1992, 1994), mtDNA (Becher and Magurran 2000), and microsatellites (Crispo *et al.*, in review; Russell 2004 and see Chapter 7) have been employed to measure gene flow in guppies in a range of Trinidadian rivers. The results are consistent; marked levels of gene flow have been detected in all cases. Crispo *et al.* (in review) found no evidence that gene flow was lower between predation regimes than within them. Nor did Crispo *et al.* uncover any reduction in gene flow in relation to the physical features, such as canopy and water velocity, that typically covary with predation risk. Divergent natural selection does not reduce gene flow (Crispo *et al.*, in review). The hypothesis of ecological speciation in guppies is not supported.

Although there is little scope for the evolution of reproductive isolation within rivers, it is conceivable that isolating barriers could develop between river systems that are physically isolated and where gene flow is prevented. This possibility is considered next.

[2] It is uncertain how guppies colonize above barrier waterfalls. Hurricanes and water spouts, translocation by birds, and populations that pre-date waterfall formation are some of the suggestions that have been advanced.

6.2 Divergence and the potential for allopatric speciation

The Caroni drainage empties westwards into the Gulf of Paria while the Oropouche river system flows east into the Altantic Ocean (Figs. 1.2 and 1.3). The watershed between the drainages lies near Valencia on the Valencia–Toco Road. The short stretch of road linking the bridge over the Valencia River (which eventually joins the Aripo River) with the bridge over the Quare River (the Quare is part of the Oropouche system) is a potent demonstration of the close physical proximity of the river systems. Indeed the drainages are separated by as little as 70 m during the wet season. In this case there is no association between genetic distance and geographic distance—guppies that inhabit these two river systems are genetically more divergent than any other populations thus far documented. Caroni guppies for instance have a greater affinity with conspecifics in Venezuela than these in the Oropouche (Fig. 1.3).

The reasons for this divergence is unclear though it is plausible that the two systems were colonized by different lineages (Boos 1984; Carvalho *et al.* 1991; Fajan and Breden 1992). Russell (2004) used the standard mitochondrial molecular clock for non-cold-tolerant fishes (2% per million years—Avise 1994) to estimate that the fish in the two drainages diverged approximately 2.5 million years ago. There are obviously large confidence intervals around this estimate. Even so it is probable, given the comparison with diversification rates in other species (Coyne and Orr 2004, table 12.1), including teleosts (McCune and Lovejoy 1998), that sufficient time has elapsed for at least some isolating barriers to emerge.

6.3 Pre-mating isolation

Although guppies have been the focus of many female choice studies, relatively few investigators have examined the relationship between genetic divergence and female preference. In the most comprehensive investigation to date Endler and Houde (1995) examined the choice behaviour of guppies collected from 11 locations in 9 rivers in Trinidad. Thirty-six combinations of localities were tested to determine both among- and within-population variation in female preference. Females from a given locality encountered native males as well as alien males from two or more localities. Overall, there was a slight—but significant ($t_{1318} = 2.74$, $P < 0.01$) preference for native over alien males. Preferences were found to vary geographically, and predation intensity appeared to influence the outcome. However, there was no consistent trend in the three comparisons that spanned the Oropouche and Caroni drainages (Fig. 6.1). Recent work in my own lab reaches a similar conclusion. Anna Ludlow and I found that although female guppies tended to make more sexual responses towards males from their own population if given the simultaneous choice of a male from another population (of equivalent predation status) within the same drainage, they do not discriminate against males from the other drainage when the choice is between these and their native population. Interestingly, female guppies also make glide responses towards *Poecilia picta* males, albeit at a lower frequency than to conspecifics. One

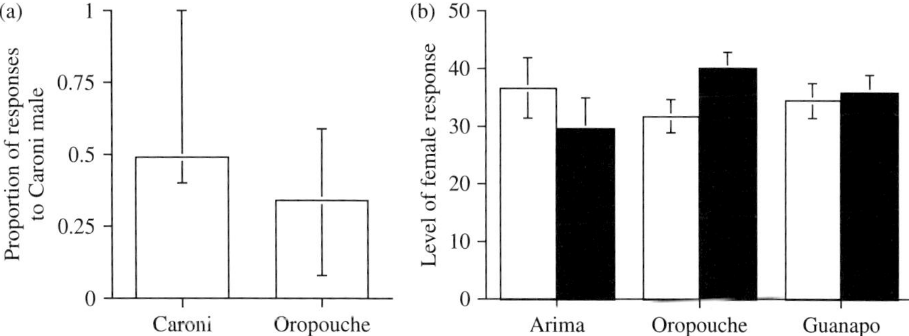

Fig. 6.1 Female guppies do not discriminate against genetically divergent males. (a) Glide responses by receptive (virgin) Tacarigua River (Caroni drainage) and Oropouche River (Oropouche drainage) females towards a Tacarigua male. Median values and quartiles. See Magurran *et al.* (1996) for details. (b) Cross drainage (Caroni–Oropouche) comparisons by Endler and Houde (1995). The origin of the receptive female guppy is indicated. In all cases she was given the choice of a male from her own population (open bars) and a foreign (other drainage—black bars) male. The mean fraction female response, ±s.e., to male displays, is shown. Endler and Houde provide details of the method. After Magurran (1998).

possible reason for these findings is that females are expressing preferences not only for native colour patterns but also for novelty (Farr 1977; Hughes *et al.* 1999). In addition, female choice tests typically make use of virgin females in order to ensure receptivity. These females are usually reared without sight of males, and invariably without direct contact with males. Early experience has been shown to influence female mating decisions (Breden *et al.* 1995; Rosenqvist and Houde 1997) and it is not inconceivable that it also affects female ability to discriminate between classes of males. For this reason it would be instructive to examine the preferences for alien versus native males in both naïve and experienced females.

Endler and Houde (1995) point out that studies of sexual selection and sexual isolation are often conducted in isolation. They argue that species recognition, and indeed reproductive isolation, may be the product of sexual selection on particular characters. Natural selection means that different traits may be the target of female choice in high- and low-predation environments. Long separation, as has occurred between Caroni and Oropouche guppies could result in divergence in the colour patterns that females prefer. Signal detection theory (Shettleworth 1998)—the detection of genuine signals against a noisy background—would be a useful tool in this type of investigation.

There are other reasons for supposing that female choice need not translate into sexual isolation. Although it has been well established that population differences in mate choice have a genetic basis (Houde and Endler 1990; Endler and Houde 1995) and that potential for further divergence exists (Brooks and Endler 2001*b*), there are features of the way in which this choice is exercised that impede the emergence of reproductive barriers. Mate choice can be divided into two components: *choosiness* and *preference functions* (Jennions and Petrie 1997). Choosiness is the effort that a female devotes to selecting a male while preference functions represent the manner in which females rank different males (see also Chapter 4). As Brooks (2002) points out, choosiness is likely to reflect the cost of choice, covarying, for example, with degree of predation risk, whereas preference functions will tend to be shaped by the processes that influence signal design, such as the physical characteristics of the environment in which mating occurs. A series of experiments using feral populations of guppies in Australia led Brooks to conclude that differences in preference function will not necessarily result in preferences for males from the native population. Instead attractive males tend to be preferred, irrespective of their origin. Moreover, females do not all agree about which males to mate with (Brooks and Endler 2001*b*), mate multiply (Kelly *et al.* 1999; Evans and Magurran 2000), switch mating partners between successive broods (Becher and Magurran 2004; Eakley and Houde 2004), opt for rare males (Hughes *et al.* 1999), and change their preference functions as they age (Kodric-Brown and Nicoletto 2001*a*). This means that when divergent populations come into secondary contact considerable intermating is likely to occur (Brooks 2002)—which is exactly what we see.

Although it is possible that discrimination against alien males is stronger than that suggested by laboratory studies, the inference of this body of work is that female preferences alone are not sufficiently strong to drive reproductive isolation among guppy populations in Trinidad.

There is one apparent exception to this pattern. The Cumaná guppy is a highly differentiated colourful morphotype of the guppy found near the coastal town of Cumaná in NE Venezuela (Alexander and Breden 2004). These fish were first collected by Franklyn Bond in 1937 and later by John Endler in 1975 (Alexander and Breden 2004), and are often called 'Endler's guppy' or 'Endler's livebearer' by hobbyists. Heather Alexander and Felix Breden argue that the Cumaná guppy is not sufficiently diverged to be considered another species of guppy. Although it has striking red and green coloration and vivid sword markings on the tail, a phylogeny based on mtDNA places the Cumaná guppy closer to the west Venezuela, east Venezuela, and west Trinidad (Caroni) guppy populations than the east Trinidad (Oropouche) guppies are to any of these. Alexander and Breden could detect no evidence of genetic incompatibility in crosses between a Cumaná population and various guppy populations. F_1s, F_2s, and back-crosses were produced and there was no significant difference in offspring number due to sire type.

Analyses of female preference tell a different story. Four virgin females were housed with four males, two of which were from the Cumaná population, with the remaining two from a geographically isolated guppy population. Females were allowed to give birth and paternity was assigned using colour patterns. The results indicated a degree of assortative mating. Over 80% of the offspring produced by Cumaná females were sired by Cumaná males whereas these males sired only around 30% of the offspring of 'regular' guppy females. This result implies that females prefer to mate with their own type of male—a result supported by dichotomous choice tests (Lindholm and Breden 2002). However, gametic incompatibilities (see below) might also contribute to the outcome. The origin of these distinctive guppies is a mystery; Alexander and Breden argue that it is unlikely that natural selection (particularly predation risk) has played a major role in their diversification and sexual isolation. There is clearly much more to be learnt about the relationship between sexual selection

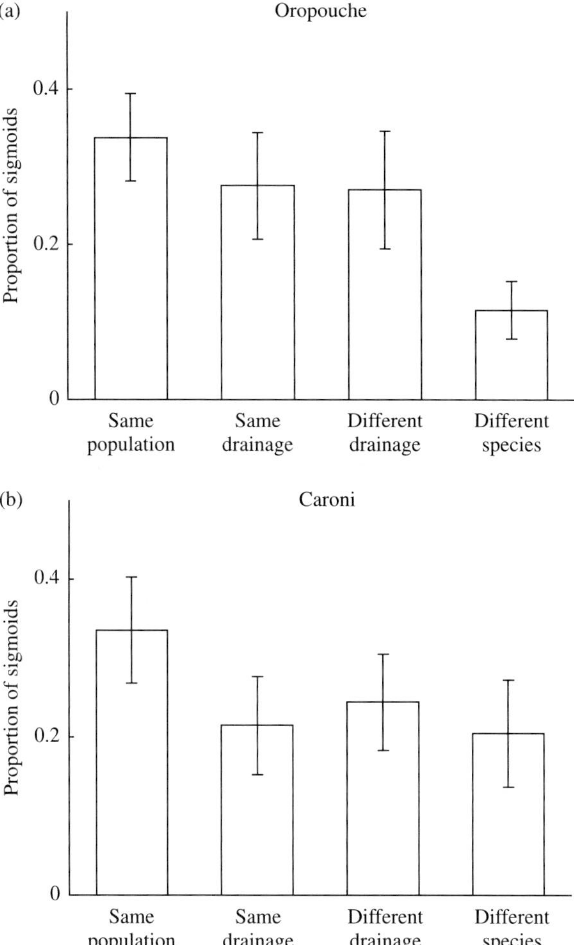

Fig. 6.2 Male choice by males of genetically divergent females. Single males were given a simultaneous choice (visual and olfactory, but no contact) of four females: same population; different population from the same drainage; different drainage; and different species. For example, an Oropouche male would be presented with an Oropouche female, a Quare female, a Lower Aripo female, and a *P. picta* female. All females were unfamiliar to males. The number of sigmoids (here presented as proportion) towards each female were recorded. The graph illustrates results from one Oropouche drainage population ((a) Lower Oropouche River) and one Caroni drainage population ((b) Lower Aripo). In both cases there is a slight preference for same population females, but no discrimination across drainages. *P. picta* females are actively courted, particularly by males of the Caroni drainage. See text for further discussion.

and reproductive isolation in this intriguing group of fish. Unfortunately, the extant populations of the Cumaná guppy appear vulnerable to anthropogenic disturbance (Felix Breden, personal communication) and we can only hope that they survive long enough to allow the details of the isolating mechanisms to be unravelled.

As Chapter 4 noted, male guppies also exert mate choice. Males, given the opportunity to court females across a gradient of genetic divergence show a slight preference for mating partners from the same population but are as likely to display to females from the other drainage as to females from a different population within their own drainage. Even heterospecifics (*P. picta*) attract attention (see Fig. 6.2 and Section 6.6. for further discussion). Heinrich and Schröder (1986) found that male *Poecilia. reticulata* did not discriminate between inbred and outbred females or females that were mutationally damaged from those that were not. This indicates that the failure of males to weight mating preferences by genetic identity may be a general phenomenon in guppies.

6.4 Sexual coercion

Males do not merely court females with the goal of soliciting consensual copulations. Females in the wild are the target of repeated sneaky mating attempts, and sperm transfer as a result of sneaky mating appears common (Pilastro and Bisazza 1999; Matthews and Magurran 2000; Evans *et al.* 2003*a* and see also Chapter 4). It is therefore likely that if two divergent guppy populations come into secondary contact, coercive mating by males will ensure that any preferences by females for males of their own population or drainage are overridden (Magurran 1996, 1998, 2001). Sexual conflict occurs when the characteristics that enhance the reproductive success of one sex are disadvantageous to the fitness of the other (Smuts and Smuts 1993; Chapman *et al.* 1995; Chapman and Partridge 1996; Gavrilets *et al.* 2001). Parker and Partridge (1998) used a game theoretic approach to examine the evolution of pre-mating isolation following secondary contact when sexual conflict is involved. A key prediction is that reproductive isolation is more likely to emerge if females are ahead in the battle of the sexes (and thus can express strong mating preferences). In contrast, in situations where males are ahead, and female choice is overridden through sexual coercion, promiscuous mating behaviour will result in gene flow that impedes the formation of mating barriers. Male behaviour is one explanation for high levels of gene flow in rivers (Russell 2004). Indeed it is of note that male choice weakens when males are competing with one another—the natural situation in the wild (see Section 6.6.). Sexual conflict thus buttresses the consequences of labile female and male choice for reproductive isolation; these need not be alternative explanations but rather jointly reduce the likelihood that pre-zygotic mating barriers will emerge during the early stages of speciation.

The potential for sexual coercion varies geographically along with predation risk (Magurran 2001). Females in high-predation localities devote more effort to predator avoidance. Males in these habitats are also less brightly coloured and are more likely to employ sneaking behaviour (see Chapter 4 for details). On the other hand, female-biased sex ratios, brighter male coloration, and the absence of severe predators suggest that the scope for sexual selection, in the form of female choice, is greater in low-predation localities. Productivity differences between upstream and downstream habitats (Chapter 2) have the potential to strengthen these patterns. Figure 6.3 presents

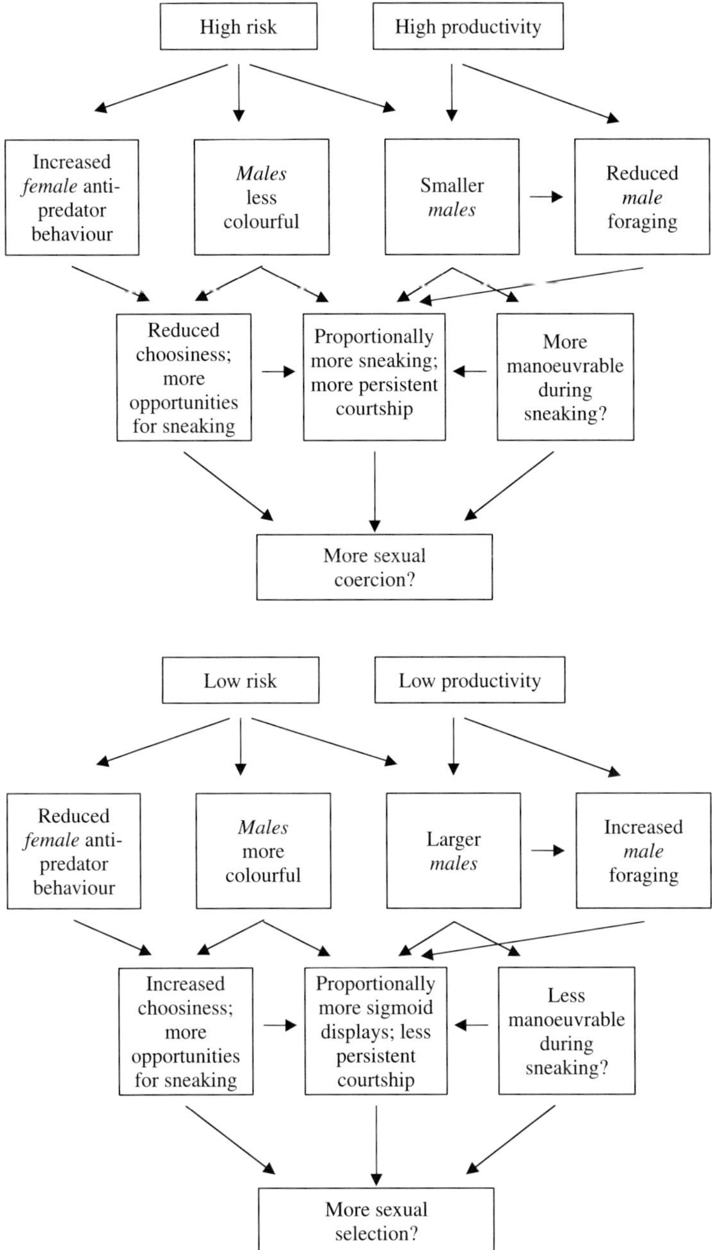

Fig. 6.3 The dual influences of predation and aquatic productivity on sexual conflict in (a) high- and (b) low-predation localities. Arrows illustrate some of the direct and indirect effects of these variables on morphology and behaviour, and highlight the scope for female choice and sexual coercion.

some of the interacting features of these habitats and their putative influences on sexual conflict. Finally, it is possible that the outcome of the battle of the sexes varies diurnally. Courtship rates decline at high light levels (Endler 1987; Reynolds and Gross 1992) particularly among larger, more preferred, males (Reynolds and Gross 1992; Reynolds *et al.* 1993) whereas sneaky mating attempts appear to be unaffected by light level (Endler 1987). Female 'victory' at dawn and dusk could thus give way to 'victory' by males at midday. Male coercion would also be muted at sites with considerable canopy cover. These are most likely to be low-predation, upstream localities—places were sexual coercion is already reduced by other factors.

6.5 Post-mating, pre-zygotic (gametic) isolation

It is increasingly appreciated that events that occur during mating itself may be important in reproductive isolation. Coyne and Orr (2004) provide an authoritative review of the topic. In brief, post-mating, pre-zygotic (or gametic) isolation covers all the barriers between copulation (or spawning) and fertilization in animals. It may be subdivided into non-competitive and competitive categories. Non-competitive isolation arises when sperm from heterospecific male are less successful at fertilizing eggs. There are a variety of ways in which this can occur. For instance, fewer sperm may be transferred during heterospecific crosses (e.g. Price *et al.* 2001), foreign gametes may be inviable in the female's reproductive tract (e.g. Gregory and Howard 1994), and fertilization may not occur when the gametes contact one another (Palumbi and Metz 1991). Competitive isolation, which is sometimes also known as conspecific sperm precedence, is manifested when females simultaneously inseminated with both heterospecific and conspecific sperm produce fewer hybrid offspring than would be expected in a non-competitive situation (Howard 1999). Coyne and Orr (2004) point out that as multiple mating is common in nature, conspecific sperm precedence could be an important reproductive barrier. Price (1997), for example, found that *Drosophila simulans* females would readily produce hybrid offspring when inseminated with sperm from *Drosophila mauritiana* males, but that few hybrids resulted when *D. mauritiana* sperm were in competition with *D. simulans* sperm. Physical displacement and incapacitation of the foreign sperm jointly accounted for the competitive advantage of the conspecific males (Price *et al.* 2000). Investigators can only be certain that competition is implicated in gametic isolation if controls using single mated heterospecifics are also carried out.

Coyne and Orr (2004) argue that gametic isolation may be historically important in speciation since these barriers could arise as a by-product of sexual selection at the molecular level. Drawing inferences from laboratory studies, which confirm that reproductive systems can evolve rapidly, they postulate that gametic isolation could be one of the fastest-evolving reproductive barriers. What then is the evidence that it might be important in guppies? The importance of post-copulatory mechanisms in determining the outcome of multiple matings underlines the potential for this type of barrier. Our own investigations (A. M. Ludlow and A. E. Magurran, unpublished data) show that *P. picta* sperm will not fertilize *P. reticulata* eggs, and vice versa, demonstrating that non-competitive gametic isolation exists between fully formed species. However, we also tested for 'conspecific' sperm precedence (or more correctly, conpopulation sperm precedence) between guppies in the Caroni and Oropouche drainages. There is no reduction, relative to within-population crosses, in the fecundity of females (of either drainage) inseminated with foreign sperm (Russell 2004). Artificial insemination (Evans *et al.* 2003*b*) is a useful technique for ensuring that females receive equal numbers of sperm (measured as number of sperm bundles) and that they are delivered simultaneously (to avoid mating order effects (Evans and Magurran 2001)). We used this approach to inseminate Caroni drainage females with sperm from both a Caroni and an Oropouche male. The reciprocal test, using an Oropouche female, was also conducted. Offspring were genotyped to assign parentage. Our results were consistent with 'conspecific' sperm precedence—sperm from the females' own drainage sire proportionally more of the offspring than would be expected by chance (Fig. 6.4). They support Coyne and Orr's prediction that gametic isolation is common among closely related taxa and that it can be one of the earliest reproductive barriers to become established.

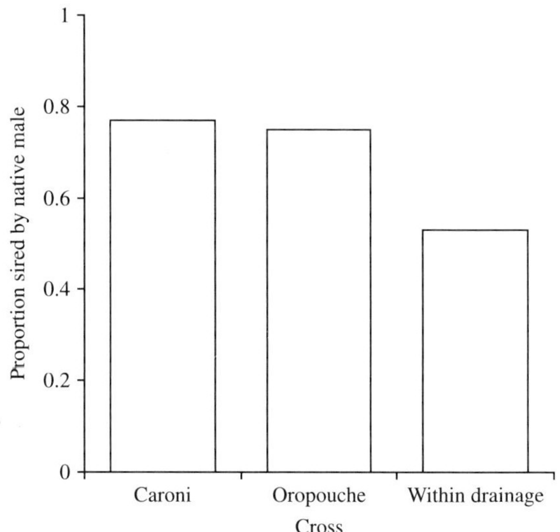

Fig. 6.4 Conspecific sperm precedence in guppies. The graph shows the proportion of a brood sired by a native male when his sperm are in competition with equal numbers of foreign (other drainage) sperm (both Caroni and Oropouche females tested). A within-drainage control is also shown. In between drainage crosses native males sire most of the brood ($P < 0.006$, 1 sample sign test, tested against expectation of 0.5). In contrast the native male has no advantage when the competition is with sperm from the same drainage ($P > 0.99$). As there is no loss of fecundity in single inter-drainage matings (Russell 2004), these results show that sperm competition is involved. Ludlow and Magurran, unpublished data.

However, freshwater fish are one of the groups most at risk. Moyle and Leidy (1992) investigated the status of these species in a range of faunas. Species were catalogued according to three overlapping categories of risk: *endangered*, that is, species where extinction appears imminent; *threatened*, that is, species that are declining and likely to become endangered; and *species of special concern*, that is, species that are in decline or with a very limited range but not facing extinction in the near future. The percentage of the total freshwater fish fauna in one of these three categories ranged from 9 to 69% across 9 regions. The median value was 28%. In North America, a well-documented region, 3 genera, 27 species, and 13 sub-species have gone extinct since 1900 (Miller *et al.* 1989). In California alone, only 31% of species can be regarded as 'secure' (Moyle and Leidy 1992). Eight out of the 14 poeciliid species native to the United States are either threatened or endangered (Johnson and Hubbs 1989).

Although the species is the usual unit of conservation effort, there is increasing appreciation that populations or sub-species, also deserve consideration (RoyalSociety 2003). As the guppy literature makes abundantly clear, populations can be the repository of considerable biological diversity. The guppy as a species is not threatened, but the guppy as a collection of natural populations is subject to many of the anthropogenic impacts that have led to a reduction in freshwater fish diversity worldwide.

The term 'evil quartet' (Diamond 1989) has come to encapsulate the problems that nudge organisms ever closer to extinction. The four impacts that species and populations face are: over-harvesting; habitat fragmentation and degradation; the presence of exotic species; and chains of extinction (see below). As the guppy is not a commercially fished species, except through the occasional provision of new material for aquarists, it is not subject to over-fishing in the conventional sense of the term. Nonetheless, it is conceivable that scientists could over-fish some populations. I review this possibility later in the chapter. The presence of exotic species, at least in Trinidad, seems to be an unlikely trigger for population extinction. However, Trinidadian rivers are increasingly being invaded by alien species that have escaped from fish farms or from home aquaria. Two species of tilapia (*Oreochromis mossambicus* and *Oreochromis niloticus*)—a notorious invasive—are already widely distributed (Phillip and Ramnarine 2001) and appear to have detrimental effects on natural communities (Magurran 2004). The congeneric *Poecilia sphenops* is another species spreading through Trinidadian rivers (Phillip and Ramnarine 2001). The danger of exotics must not therefore be entirely discounted. Invasions by foreign guppy populations, as a result of experimental manipulations, can have far-reaching consequences for native genotypes (see Section 7.7) even if they are unlikely to result in extinction in the classic sense. Chains of extinction, in which the decline or extinction of one species causes the decline or extinction of others are also unlikely to apply to guppy populations. There is no evidence that I know of to suggest that removal of other species from a community will adversely affect guppies, even though the event will change the pattern of selection. It is the second member of Diamond's quartet, habitat fragmentation and degradation, which gives most cause for concern.

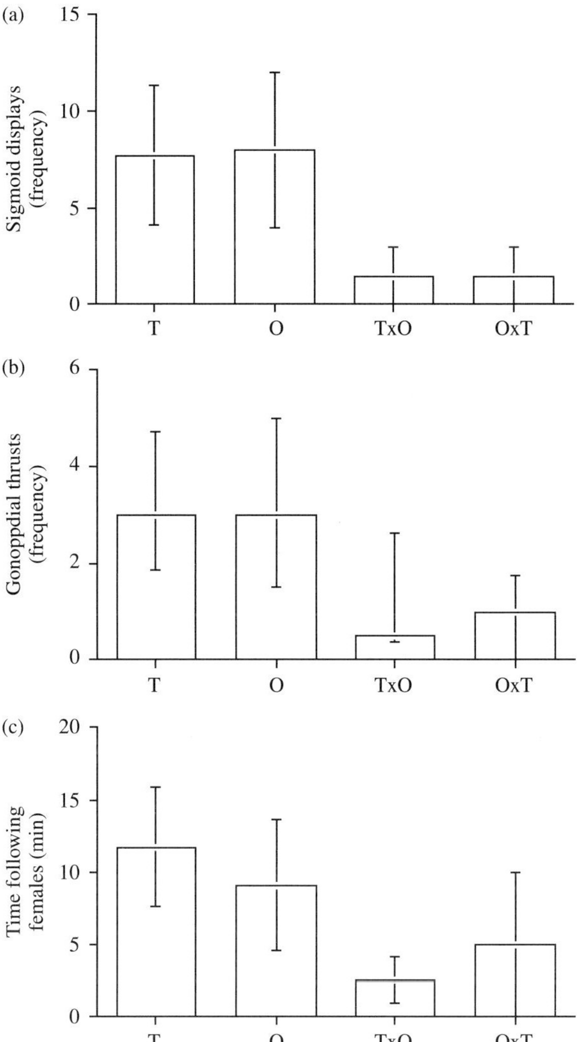

Fig. 6.5 Mating behaviour by males in the parental and reciprocal F_1 lines. T = Tacarigua River (Caroni drainage) and O = Oropouche River (Oropouche drainage). (a) Frequency of displays, (b) frequency of gonopodial thrusts, and (c) time spent following females. Median and quartile range shown in all cases. Trials lasted 15 min. After Russell (2004).

likely to experience inviability or sterility. Coyne and Orr (1989, 1997) confirmed that this pattern holds among young taxa in *Drosophila*. Hybrid problems afflicting both sexes tend to occur later on in the speciation process (Coyne and Orr 2004). Yet another way in which isolation between Oropouche and Caroni guppies matches the general pattern for animals is that hybrid sterility is evolving more quickly than hybrid

6.6 Post-zygotic isolation

The long separation between Caroni and Oropouche guppies also suggests that it might be productive to search for post-zygotic isolation. Post-zygotic isolation can be expressed in two forms (Coyne and Orr 2004). 'Extrinsic' isolation, such as ecologically mediated isolation, occurs when hybrids are at a disadvantage in a given habitat due to their intermediate phenotype. The reduced growth rate of F_1 hybrids of the limnetic and benthic three-spined stickleback (*Gasterosteus aculeatus*) morphs found in Paxton Lake, in British Columbia, is an example of this phenomenon (Hatfield and Schluter 1999; Rundle 2002). However, since the Caroni and Oropouche drainages offer parallel environments, notably in terms of predation regime, it is unlikely that hybrids formed by crossing populations experiencing similar levels of risk will be disadvantaged in any way. Nor does it seem plausible that the disadvantages experienced by hybrids produced by inter-drainage, high-predation, low-predation crosses, will be any greater than those resulting from crosses between predation regimes within a drainage. Behavioural sterility, another form of extrinsic isolation, results when hybrids are unable to find mates. Given the preferences that both male and female guppies display for novel partners, this form of extrinsic isolation also seems improbable in guppies. 'Intrinsic' isolation, on the other hand, is developmentally mediated and expressed by a reduction in hybrid fertility and viability. Hybrids may either be inviable or sterile. Sterility can be both physiological, manifested as developmental defects in the reproductive system, or behavioural, such that hybrids are fully or partially incapable of courtship (Coyne and Orr 2004). (As Coyne and Orr point out, the distinction between extrinsic and intrinsic behavioural sterility is that hybrids have intermediate courtship patterns in the extrinsic form, whereas behaviour is disrupted during intrinsic sterility.) Coyne and Orr provide an excellent account of the evolution of sterility, which can arise as a by-product of genomes that are geographically isolated.

At first glance guppy population crosses might not seem a very fruitful source of intrinsic post-zygotic isolation. Mixed population tanks in the laboratory typically support thriving stocks, and crosses between populations generate fish capable of the full courtship repertoire. However, mindful of the long separation of the Caroni and Oropouche drainages we (S. T. Russell and A. E. Magurran) had a more careful look. Multiple aspects of reproductive behaviour and output were examined in crosses (F_1, F_2, and back-crosses) that had been produced using fish from the Tacarigua River (Caroni drainage) and Oropouche River (Oropouche drainage) (Russell 2004). Around 80% of attempted crosses yielded progeny. There was no diminution in brood size in F_1s relative to the parental stock, but fewer offspring were produced, relative to female body size, in F_2 and various backcross generations. Sperm counts were also reduced in the F_2 cross. Significant male behavioural sterility was uncovered in the form of a reduction in the number of sigmoid displays, in the frequency of sneaky mating attempts, and in the time devoted to courtship in F_1 hybrids (both reciprocal crosses) relative to parental stocks (see Fig. 6.5).

These experiments suggest that guppies follow Haldane's rule (Haldane 1922), which states that the heterogametic sex (i.e. XY males in the case of guppies) are more inviability. And because the work employed conspecific populations as opposed to distinct species the results are of broader interest as they demonstrate that behavioural dysfunction may lead to speciation.

6.7 Learned mate recognition and reproductive isolation

Learned mate discrimination, which can help reinforce barriers between emergent taxa, has been documented in a wide range of animals including fruit flies (Kim *et al.* 1996; Kim and Ehrman 1998) and birds (Lorenz 1937; Clayton 1988; Oetting *et al.* 1995; Slagsvold *et al.* 2002). Guppies provide novel insights into its operation and contribution to reproductive isolation. Male guppies pursue females relentlessly and switch to homosexual behaviour if housed in single sex groups. Males even direct frequent courtship displays and sneaky matings towards any heterospecific females they encounter during secondary contact. However, Caryl and Edna Haskins (Haskins and Haskins 1949) made the interesting discovery that after some days male guppies begin to distinguish between guppy females and other poeciliid females. The Haskinses further reported that discrimination of the correct females occurs with a high degree of accuracy after the fish are together for about 20 days. Magurran and Ramnarine (2004) followed these ideas up in an experiment that allowed two male guppies to compete for two females, one a guppy female, the other a *P. picta* female. Males from sites where guppies are the only poeciliid present initially courted both females with equal enthusiasm, but within a week they were directing most of their attention towards the conspecific. Indeed, these allopatric males acquired a level of discrimination that matched that of males from localities where *P. picta* and *P. reticulata* occur sympatrically (Fig. 6.6).

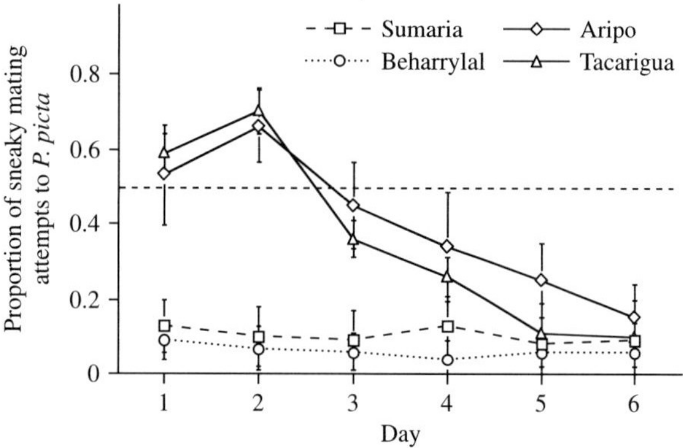

Fig. 6.6 Learned mate discrimination. Proportion (mean ± 95% confidence limits) of sneaky mating attempts that a focal guppy male directed towards a *P. picta* female over successive days. Males were in competition and had a choice of a *P. picta* or *P. reticulata* female. The dashed line represents random choice. Four populations were studied. Sumaria and Beharrylal guppies occur sympatrically with *P. picta*, whereas Aripo and Tacarigua guppies are in allopatry. All fish were wild caught. See Magurran and Ramnarine (2004) for further details.

It seems unlikely that learned discrimination of females is a form of familiarity learning since guppies need to associate for a period of around 12 days before they can distinguish familiar conspecifics (Griffiths and Magurran 1997*a*). And, since the females in our (Magurran and Ramnarine 2004) experiment were non-receptive, a sexual response by the female is not a necessary condition for learning to take place. Nonetheless, it appears that female behaviour does play a role and receptivity may accelerate the learning process. *P. picta* females respond to persistent courtship from their own males by abruptly ceasing to swim—a reaction that resembles the behaviour of a guppy female just as she is about to glide towards a male and initiate a consensual copulation. Naïve male guppies are initially fooled by this behaviour (Fig. 6.6) and make repeated sneaky mating attempts towards these stationery *P. picta* females, which they pursue in preference to conspecifics. One reason for their subsequent switch to guppy females may be the failure of *P. picta* females to complete the glide, arch, and copulation sequence that a receptive conspecific would engage in Liley (1966). Another reason could be that *P. picta* gonopores are smaller in diameter than the *P. reticulata* gonopores, and the act of attempted copulation provides confusing tactile feedback. Observation of the behaviour of other males may also contribute to the learning process. It would be interesting to ascertain if social learning is involved and whether males paired with experienced tutors do achieve accurate mate discrimination more quickly.

Liley (1966) found that *P. reticulata* males reared with *P. picta* females initially courted the heterospecific females from the point of first gonopodial development, but soon began to display to and court females of their own species. This experiment and others in which males were separated from females for varying lengths of time led Liley to propose that there may be a sensitive period during development when *P. reticulata* males learn to discriminate males and females and possibly even to distinguish between females. Similarly, exposure to certain male phenotypes shapes the mating choices of adult female guppies (Breden *et al.* 1995; Rosenqvist and Houde 1997; Jirotkul 1999*a*). Early experience can thus influence mating preferences in guppies. However, the acquisition of sexual preferences need not be a single process (Shettleworth 1998). Many bird species show a predisposition to recognize conspecific mates (Immelmann 1972). These preferences are moulded by early events, including filial imprinting, but are consolidated by sexual contact later in life (Bischof 1994). Indeed Irwin and Price (1999) argue that because the benefits of recognizing particular individuals persist through life, the ability to learn to recognize the traits of others must also remain. The guppy system confirms that learned mate discrimination can continue through life. The data also support the supposition that learned discrimination of species is underpinned by the same mechanisms that produce learned discrimination of conspecifics (Ryan and Rand 1993).

One interesting aspect of the learned mate recognition experiment (Magurran and Ramnarine 2004) is the strong preference for conspecific females exhibited by males collected from sympatric populations, that is, localities where both poeciliids naturally occur. This could be interpreted as evidence for geographic variation in reinforcement, and indeed this might well be the conclusion had the responses of wild-caught males been tested on a single occasion. However, as the wild males have different opportunities to learn, the varying responses of the wild fish are not, in themselves, an indication that reinforcement varies adaptively. Fortunately, I can report that naïve laboratory-reared fish from the four populations investigated by Magurran and Ramnarine (2004) vary in their preferences for *P. picta* females when they first encounter them in a competitive choice test. Males sired by parents from sympatric sites prefer conspecific females while those from allopatric localities mate at random.

Although male guppies that occur sympatrically with *P. picta* show a strong and consistent preference for conspecifics, they sometimes attempt matings with heterospecific females. Molecular methods reveal that approximately 5% of *P. picta* females in these habitats have *P. reticulata* sperm in their gonopores (Russell *et al.* 2005). (A similar proportion of *P. reticulata* gonopores contain *P. picta* sperm.)

6.8 Relative importance of different reproductive barriers

These investigations demonstrate that isolating barriers are becoming established among guppy populations, and shed light on the relative importance of the different mechanisms, as well as on the order in which they arise. New data suggest that post-mating pre-zygotic isolation and post-zygotic isolation are most important in the early stages of speciation in this system. Despite the wealth of studies demonstrating that both males and females exert mate choice, marked individual differences in the manner in which females choose partners, combined with multiple mating and promiscuity in males, imply that reproductive isolation is unlikely to arise as a by-product of assortative mating. This does not mean that mate choice cannot reinforce barriers between well-established taxa. Indeed, Liley's classic (1966) monograph illustrates well the plethora of behavioural and morphological differences between congeneric poeciliids. But it does suggest that sexual isolation will not provide the first step along the road to speciation. The interesting case of the Cumaná guppy provides an intriguing counterpoint to these conclusions. Here, pre-mating isolation appears to be important and seems to have been one of the earliest barriers to emerge. Although gametic isolation between Cumaná and Trinidadian guppies cannot be entirely ruled out, there appear to be some fundamental differences between the two types of poeciliid that would warrant further investigation.

Coyne and Orr (2004) observe that multiple barriers between emerging species are to be expected given that speciation is typically a long process. Gametic isolation and post-zygotic isolation both operate during matings between Caroni and Oropouche guppies. We cannot tell which of these barriers arose first, but this does not matter as isolation between the drainages is not yet complete. Nor indeed is it possible to predict whether two robustly differentiated guppy species will eventually emerge. Coyne and Orr further note that the fact that post-zygotic barriers act late in the life cycle does not mean that they are unimportant in speciation. Indeed, they give nine reasons why these barriers may be significant. Two are particularly pertinent here.

First, in contrast to other forms of isolation, intrinsic post-zygotic isolation is likely to be permanent. Thus, genetic differences between populations that lead to sterility in hybrids are likely to be perpetuated once established. Second, if the genes responsible for intrinsic post-zygotic isolation are to some extent recessive in the F_1 generation, but expressed more strongly in the F_2 and backcross generations—as appears to be the case in guppies—post-zygotic isolation will form an even stronger barrier in successive generations of crosses.

A further reason for concluding that barriers that operate during and after mating are more important in the guppy system is that the factors that impede the evolution of pre-mating isolation are not expected to weaken over time.

Much of the foregoing discussion explores the way in which potential reproductive barriers would operate if tested when divergent populations meet. But how likely is it that secondary contact will occur? Guppies in the Caroni and Oropouche drainages have been separated for a long period of time. Nonetheless, they may encounter one another, not only in the laboratory but also in the wild. Anthropogenic impacts on freshwater systems in Trinidad are one way in which divergent populations of guppies might meet. Casual translocations of the sort that occur when small children collect fish as pets are one possibility. A more likely cause is a deliberate introduction by scientists. The first guppy transplant in Trinidad (Magurran *et al.* 1992; Shaw *et al.* 1992, and see Chapter 7) breached the divide, unbeknownst to its instigator, Caryl Haskins, who was unaware of the genetic structuring of populations. Natural changes in geomorphology and stream flow are a second reason why fish in the Caroni and Oropouche systems might encounter one another. As noted at the beginning of the chapter, the two drainages abut one another and it would take only a small amount of flooding, or one of the earthquakes to which Trinidad is prone, to reconfigure the landscape.

6.9 Sympatric speciation

With a handful of notable exceptions, such as the sympatric morphs of arctic charr (*Salvelinus alpinus*) and sticklebacks (*G. aculeatus*) in Iceland, there are few compelling examples of sympatric speciation in fish and other vertebrates (Coyne and Price 2000; Coyne and Orr 2004). Most of the plausible examples of sympatric speciation are related to differences in foraging behaviour and morphology. For example, Lake Thingvallvatn in Iceland supports four charr morphs that vary markedly in diet, morphology, and habitat use (Skúlason *et al.* 1999). Guppies provide little scope for speciation driven by trophic polymorphism as males and females occupy different feeding niches (Fig. 3.8). Indeed the long time period required for the emergence of post-mating mechanisms under allopatry testifies to the unlikelihood of reproductive barriers forming in sympatry. The case of the Cumaná guppy is the sole example where incompletely diverged forms of *P. reticulata* persist in sympatry (Alexander and Breden 2004). Nonetheless, it is probable that the initial divergence from Trinidadian guppies took place in allopatry.

6.10 Conclusions

Contrary to first impressions, reproductive barriers are beginning to become established among genetically divergent guppy populations. These occur in the post-mating arena, both as gametic isolation and as post-zygotic intrinsic isolation in the form of male behavioural sterility. There further appears to be hybrid breakdown in embryo viability, brood size, and sperm number in the F_2 and backcross generations.

Although behavioural (pre-mating) isolation may evolve more rapidly than post-mating isolation among allopatric populations of species which are undergoing strong sexual selection (Mendelson 2003), there are no *a priori* reasons for predicting that this should be a general pattern (Coyne and Orr 2004), especially since sexual selection can lead to gametic isolation and sterility (Turelli *et al.* 2001). Furthermore, the characteristics of both male and female mating behaviour will limit the extent of assortative mating, while parallelism in ecological conditions in the different drainages will select for similar choice criteria. The guppy system shows how two or more barriers can become established in tandem during the early stages of speciation and illustrates why it may not be meaningful to search for a single isolating mechanism that triggers speciation. We cannot tell whether the barriers that presently exist will be the ones that eventually ensure complete isolation, but it seems reasonable to predict that this will be the case assuming that there is no large-scale intermixing of Caroni and Oropouche populations. There may also be highly differentiated populations elsewhere that would provide independent tests of the patterns observed here.

In contrast to other taxa, notably *Drosophila* (Coyne and Orr 2004), little is known about the genetic architecture of reproductive barriers in guppies. This type of research has been impeded by the absence of a genetic map for the species. Anticipated advances in this field mean that it should soon be possible to address key questions, including the number of loci that play a role in the early stages of speciation.

7

Conserving a natural experiment

One of the most pressing concerns for evolutionary ecologists is the rate at which species are being lost as a consequence of human impacts on the planet. It is unclear exactly how many species are at risk; the IUCNs definitive list of species loss (www.redlist.org) is probably a substantial underestimate (May 2002). However, most commentators agree that the situation is serious and that species extinctions are likely to increase. Freshwater fish are an especially vulnerable taxon (Warren *et al.* 2000; RoyalSociety 2003) as they are directly affected by our society's growing need for water and its proclivity for modifying and damaging aquatic habitats and introducing alien species into them.

There are three different ways in which guppies stimulate the interest of conservation biologists. First, there is an urgent need for model systems that can be used to experimentally examine the multiple and synergistic consequences of anthropogenic activities. Guppies provide a tractable system that can, for example, help elucidate the effects of small population size, loss of genetic diversity, and response to deteriorating water quality. A related topical issue is the manner in which 'contemporary' or 'rapid' evolution influences the response to factors, such as environmental change and exotic species (Stockwell *et al.* 2003). The guppy, as an exemplar of rapid evolution in the wild, is a natural choice for experimental investigations of evolutionary change linked to human impacts. Second, although the guppy, as a species, is in no danger of extinction, the guppy as a collection of divergent populations—Haskins's 'natural experiment'—is at risk. Native populations, particularly the landmark guppy populations in Trinidad's Northern Range, are subject to habitat deterioration, pollution, and possibly even over-harvesting by scientists. Artificial introductions have proved very informative but may lead to irreversible changes in a river. The benefits of these manipulations need to be set against the costs. Finally, the guppy is itself an invasive species in the many habitats into which it has been released either deliberately in an attempt to control mosquitoes, or accidentally through the escape of pet fish.

7.1 Guppies as a model species for conservation

A matter of considerable interest to conservation biologists is the relationship between molecular markers and adaptive traits. It is widely assumed that levels and distribution of genetic diversity are linked in some meaningful way to fitness variation in ecologically significant attributes. This follows from Fisher's (1930, 1958) 'fundamental theorem' of natural selection, which predicts a correlation between adaptive evolutionary change and genetic diversity (Carvalho *et al.* 2003). However, the relationship is not necessarily straightforward as adaptive characters are usually under polygenic control (Reed and Frankham 2001). Moreover, molecular variation and complex traits may respond in quite different ways to evolutionary history and selection (Hedrick 2001), and while the phenotypic expression of a character is typically environment-dependent, most molecular markers are not. Nonetheless, investigators are increasingly attempting to relate molecular variation and quantitative genetic variation (see, e.g., Lynch *et al.* 1999; Kruuk *et al.* 2000). One important function of this approach is to ascertain whether populations or species in which molecular diversity has been eroded, have also lost adaptive variation.

'Adaptive genetic variation' can be defined as 'genetic variation that is correlated with variation in lifetime or total fitness of individuals (Endler 2000, p. 251). Heritable variation exists in a wide range of traits encompassing life history, behavioural, morphological, and physiological characters. Typically there are differences in average heritabilities (h^2) among classes of traits (Roff and Mousseau 1987; Hoffmann 2000). In *Drosophila*, for instance, morphological and physiological traits have higher heritabilities than life-history and behavioural traits (Roff and Mousseau 1987). The same is probably true for guppies. For example, male secondary sexual characteristics often have high values of h^2 (Brooks and Endler 2001*a*; Karino and Haijima 2001) whereas the heritability of behaviours, such as shoaling is generally low (Paxton 1994). A comprehensive analysis is, however, awaited.

Powerful new approaches to quantitative genetics investigations, such as the North Carolina Designs (Lynch and Walsh 1998), are efficient at segregating the genetic and environmental components of phenotypic traits. One common drawback is that large sample sizes are required to deliver reliable estimates of genetic variance components and heritabilities. This imposes logistical and sometimes ethical constraints, particularly for vertebrates, which are the usual focus of conservation concern. (69% of papers in the conservation literature are devoted to vertebrates even though the group accounts for a mere 3% of species in nature (Clark and May 2002).) Fortunately, the characteristics of the guppy that have promoted its popularity as an aquarium fish lend themselves to this type of work. A recent investigation involving four populations of guppies (Oosterhout *et al.*, unpublished data) uncovered a consistent positive correlation between molecular and quantitative genetic variation. Interestingly, the Upper Aripo population, which is relatively depauperate in terms of its molecular variation (Carvalho *et al.* 1991), had significantly lower heritabilities for a range of life-history and other traits. Equally intriguing is the observation that some characters, notably male sexual vigour, were not related to loss of genetic diversity suggesting that sexual selection may help counteract the effects of inbreeding or evolutionary history. An ongoing multi-generational experiment, in which founder population size and mating system are being manipulated, is confirming that sexual selection is instrumental in maintaining genetic diversity in guppy populations (Gunilla Rosenqvist, personal communication).

7.1.1 Inbreeding

One of the paradoxes of the guppy system is that while every male in the wild has a different colour pattern, inbred lines, with identical male coloration, can be readily produced and sustained in the laboratory or home aquaria. Some of the best known of these strains, such as *Maculatus*, are almost 90 years old (Schmidt 1920; Winge 1922*a*; Winge and Ditlevsen 1947, and see Chapter 1) yet continue to thrive, and are easy to maintain (personal observation). These strains have been employed as useful colour markers, and were particularly useful in paternity assignment in the days before molecular techniques were developed (e.g. Farr 1977). There have been relatively few investigations of their behaviour *per se*. Farr (1976), however, noted differences in the courtship of four strains of guppies, with the *Pauper* strain exhibiting the lowest level of sexual behaviour. Although the various components of Farr's 1976 experiment are not directly comparable, as he tested fish in both the presence and absence of male–male competition, it appears that the sexual vigour of these inbred lines is generally similar to that of a polymorphic 'stock' strain. In fact *Maculatus* males are characterized by their high rate of courtship display (Farr 1980*a*). There are no reported strain differences in agonistic behaviour (Farr 1976).

Behaviour, then, seemingly does not vary in any systematic way across strains. Sex ratios, in contrast, are more likely to be female biased in older lines. Farr (1981) estimated baseline sex ratios in seven strains that ranged from *Maculatus*, the oldest, to *Nigrocaudatus/Filigran* and *Blau Iridescens*, which were first described by Dzwillo (1959)—see Fig. 7.1. Farr attributes the outcome to Y-linked genes that either lead to a decease in the production of Y-bearing sperm or reduce the competitiveness of Y-bearing sperm for ova. He is not convinced that inbreeding, or local mate competition, can account for this, arguing instead that the deterioration of the Y chromosome through the accumulation of deleterious alleles (Nei 1970; Charlesworth 1978) may be responsible. Brooks (2000) uncovered another consequence of the accumulation of deleterious alleles on the guppy Y chromosome when he documented a strong negative genetic correlation between male attractiveness and both offspring survival and number of sons maturing. Brooks reminds us that colour pattern genes and deleterious alleles can be tightly linked on the Y chromosome (Haskins *et al.* 1970). An unintended by-product of artificial selection for Y-linked colour patterns could thus be the generation of strains that become progressively female biased.

The very existence of multiple monomorphic strains of guppies might imply that the species is relatively protected against inbreeding problems. Nonetheless, detectable phenotypic changes are manifested after even modest amounts of inbreeding. Sheridan and Pomiankowski (1997*b*) conducted sib-matings for 1–2 generations on guppies from the Paria and Aripo Rivers. Inbreeding depression was recorded for both, area of coloration and number of colour spots. Oosterhout *et al.* (2003*b*) performed a more extensive analysis of guppies from two high-predation populations in which inbreeding was continued for three generations. A significant reduction in both black and orange coloration and in male sexual behaviour was observed. Indeed,

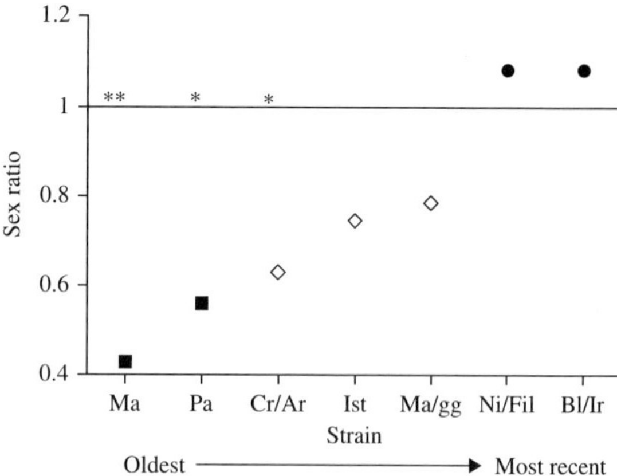

Fig. 7.1 Relationship between sex ratio and strain age. Farr (1981) determined the baseline sex ratio of seven guppy strains by raising at least 50 guppies per strain in small laboratory aquaria, and assigning sex using Y-linked colour markers. The strains were Maculatus (Ma), Pauper (Pa), Cream/Armatus (Cr/Ar), Istanbul (Ist), Maculatus/Gold (Ma/gg), Nigrocaudatus/Filigran (Ni/Fil), Blau Iridescens (Bl/Ir), and are arranged from left to right in approximate order of age. Maculatus and Pauper are the oldest and were first described by Schmidt (1920) and Winge (1927), respectively. The most recent strains in the analysis, Ni/Fil and Bl/Ir, are attributed to Dzwillo (1959). Other strains are intermediate in age. Sex ratios are presented as proportion of males. Ratios that are significantly different from unity (indicated by line through 1) are denoted as $^{**}P < 0.01$ or $^{*}P < 0.05$. Data are taken from table 1 in Farr (1981).

up to 25% of the variation in colour pattern across generations was attributable to inbreeding. Nakadate *et al.* (2003) detected inbreeding depression in survival and salinity tolerance after a single generation of full-sib mating (see also Shikano *et al.* 2001*a, b*). It is possible that preferences for unfamiliar mates (Hughes *et al.* 1999; Kelley *et al.* 1999), and partner switching by females (Becher and Magurran 2004; Eakley and Houde 2004) have evolved to facilitate outbreeding. This natural polygyny and polyandry make it unlikely that wild guppy populations will experience inbreeding depression, but it is something that should be borne in mind, should it ever become necessary to maintain fish in captivity for re-stocking purposes.

7.2 Populations in peril

The erosion of biological diversity worldwide is a matter of great concern (RoyalSociety 2003). The situation is so grave that the 2002 World Summit on Sustainable Development set as one of its targets, a significant reduction in the current rate of loss of biological diversity by 2010. Conservation effort is usually focussed on charismatic species, such as pandas and tigers, which readily attract public sympathy. The impression given is that large terrestrial mammals are especially vulnerable.

Researchers who have been visiting Trinidad for many years report a reduction in habitat quality in many Northern Range rivers (Endler 1986 unpublished report; Seghers 1992). The causes are manifold and by no means unique to Trinidad. Erosion and flash floods are exacerbated by logging and 'slash and burn' agriculture. Quarrying leads to siltation. Rivers are on the one hand the source of water, and on the other the receptacle of industrial pollution, agricultural waste, and sewage. The pressure on the system is evident from the statistic that water demand in Trinidad and Tobago rose from 297 million m^3 per annum in 1997 to 336 million m^3 per annum in 2000 (Water Resources Management Unit 2002). Population growth is a major contributory factor. The population of Trinidad and Tobago remained at around 20,000 between 1500 and 1800 (Caldwell 1995). By 1900, it had increased to 270,000. This more than doubled to 640,000 by 1950. Today the population is approximately 1.3 million.

Pollution is now recognized as an increasingly serious problem in Trinidad. For example, illicit dumping of lead wastes has contaminated rivers and wetlands, and caused acute lead poisoning in children (Mohammed *et al.* 1996). Dawn Phillip (1998) undertook a comprehensive survey of fish communities and the water quality of the habitat in which they were found, in Trinidad and Tobago. Her analysis revealed that most streams in Trinidad, and a few in Tobago, were perturbed. The most severely polluted rivers are located in the west of Trinidad, the predominant urban area, and in the SW peninsula, where the oil industry is based. There is a marked reduction in the diversity of polluted fish communities (Phillip 1998; Magurran and Phillip 2001*b*) and in some cases stretches of rivers are entirely devoid of fish. Interestingly, the last species to persist in heavily polluted sites are *Poecilia reticulata* and *Rivulus hartii.* This convergence with pristine 'upstream' assemblages illustrates why diversity statistics need to be used with caution in environmental assessment (Magurran and Phillip 2001*b*).

Extirpation of fish species, eventually even guppies, is the most obvious consequence of pollution. However, sub-lethal effects are beginning to receive more attention. Endocrine disrupting chemicals, such as phthalates, are present in Trinidadian rivers (Moore and Karasek 1984). Haubruge *et al.* (2000) found a significant decline in sperm number (of between 40% and 75%) in guppies exposed to the xenobiotics tributyltin and bisphenol A for 21 days. These authors suggest that spermatogenesis may be inhibited by interference with Sertoli-cell function. Erik Baatrup and his colleagues have examined the effects of a range of endocrine disruptors including fungicides and pesticides, and have documented multiple consequences for reproductive function including disruption of male courtship behaviour, reduction in sperm count, reduced fecundity, and a female bias in the sex ratio of offspring (Bayley *et al.* 1999, 2002, 2003; Baatrup and Junge 2001; Toft and Baatrup 2001, 2003). Exposure of males to the anti-androgenic fungicide vinclozolin is sufficient to cause a reduction in female brood size (Bayley *et al.* 2003).

Trinidad and Tobago was a pioneer in watershed conservation. The Main Ridge of Tobago became the first forest reserve in the Western Hemisphere in 1765 when it was set aside 'for the protection of the rains'. The country is well endowed with forward-looking environmental legislation and policies though lack of coordination between government agencies has impeded conservation efforts (Seghers 1992). The National Water Resources Management Policy is intended to 'unify all of these various initiatives and provide a strong direction and vision for the effective management of the nation's water resources in an integrated and sustainable manner' (Water Resources Management Unit 2002). It is to be hoped that they can achieve this worthy goal.

7.3 Distribution of research effort in the Northern Range in Trinidad

The increased interest in the Trinidadian guppy system inevitably means that more scientists are visiting Trinidad and working there. And, as a result of background information on and familiarity with certain places, some sites tend to be visited more often than others. In an attempt to evaluate the impacts that we as scientists might have, I searched the literature for reports of collections of guppies or other types of manipulation, such as mark-recapture work, that have the potential to impact natural populations. Fig. 7.2 illustrates my findings. Although I have endeavoured to be comprehensive it is not always possible to get a clear picture of the nature of the impact that a study might entail. For example, some investigators report that guppies are

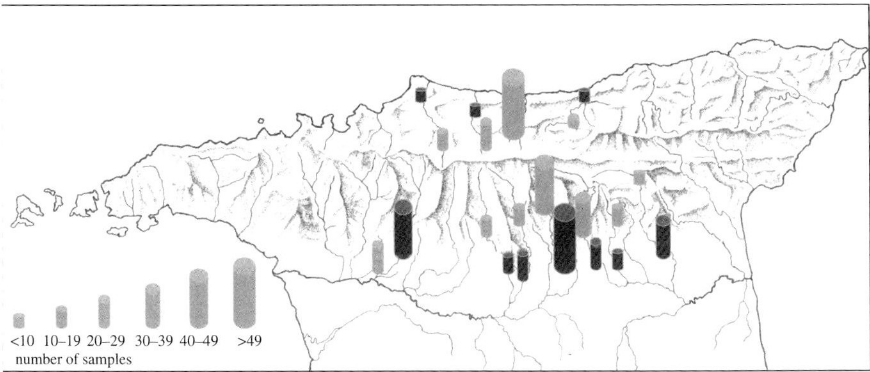

Fig. 7.2 Distribution of research effort in the Northern Range in Trinidad. Rivers are divided into high-predation (black columns) and low-predation (grey columns) sections. For clarity, data for the different tributaries within a river are combined—the El Cedro, for example, is treated as part of the Guanapo. Each 'sample' represents one reported collection or field manipulation of guppies. In some cases collected fish were returned to the river, for instance as part of a mark-recapture survey. The size of a column indicates the number of samples taken since 1990. Rivers that were sampled only once, or sites whose locations I was unsure of, are excluded. River names can be found in Fig. 1.3. As this map reveals, some sites are favoured by guppy researchers over others. See text for further discussion of the method, and the biases associated with it.

collected, examined, and returned to the site of collection. In other cases researchers have failed to record the fact that they did release guppies where they had caught them—I plead guilty here. There are no mentions of guppies being released in localities other than the one they were collected from, though this probably happens from time to time. I have excluded accounts of observation without sampling from my tally of samples, though arguably even the presence of researchers in an area causes some disturbance.

My census begins in 1990 and concludes in August 2004. It is restricted to peer-reviewed papers and to Northern Range rivers. Reviews are not counted unless they incorporate data not published elsewhere. The census date refers to the date of publication since sampling date is not always given. I am aware that there are considerable numbers of theses—batchelors, masters, and doctoral—focussed on Trinidadian guppies but no single list of these exists and pre-doctoral dissertations in particular can be hard to get hold of. I have also excluded accounts in the 'grey' literature or in popular magazines, again because of the difficulty of conducting a comprehensive survey, even though fish collections are also reported here. Nor was I able to include investigations that acted as pilot studies for later work, or were never completed, or were rejected by journals, as well as instances where fish were collected but died or were discarded. Some published descriptions of collection sites are so vague as to be unusable and a few papers describe their fish merely as 'descended from Trinidadian stock'. Against this, there is bound to be some double counting since there appear to be cases of several papers using fish descended from the same stock. Even when this is factored in, it is likely that the magnitude of the recorded research effort depicted in Figs 7.2 and 7.3 is a considerable underestimation of its true incidence. On the other hand, there is no reason to suppose that the documented distribution of research effort across river systems is systematically biased.

Each 'sample' represents a reported collection or manipulation. The distribution of collection size—if it is reported at all—is discussed in the following section.

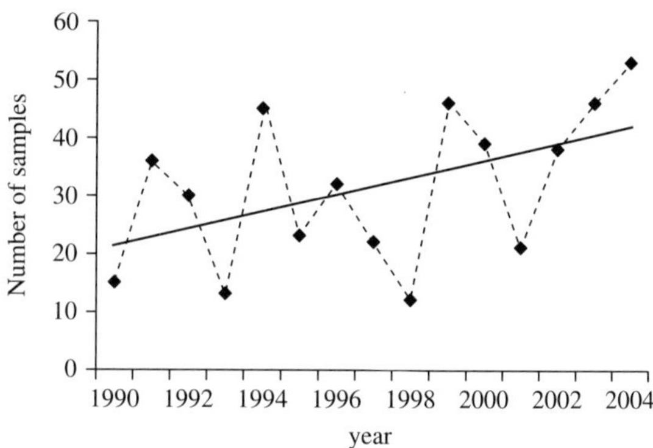

Fig. 7.3 Number of 'samples' of guppies collected in Trinidad since 1990, by year of publication. A trend line is also shown.

Where there have been 6 samples of, say, 50 fish, spread along a river section, I count this as 6 records, not 1. For clarity, I divide each river into high-predation and low-predation sections. Rivers may include one or more tributaries. For example, the Naranjo is counted as part of the Aripo River and the El Cedro as part of the Guanapo. There is no implication, therefore, that all of the collections in a river section have been taken at the same place. Nevertheless, collection sites will not be randomly dispersed or evenly distributed along a river since access is an important consideration when sampling. Many sampling spots are close to roads or popular picnic sites and individual research groups tend to visit the same localities year after year. Some sites are favoured because the contrast between high- and low-predation regimes is particularly crisp. Moreover, guppies are not equally abundant in all parts of a river. They tend to be rather sparse in shallow riffles or zones of fast flow, and will avoid deeper pools when predators are present. And because the dry season is most conducive to field work much of this research is concentrated in a fairly short period of the year. The Easter or spring break, which typically falls towards the end of the dry season, is an optimal field season for many.

The picture that emerges (Fig. 7.2) is one of concentrated research effort in a relatively few rivers. The Paria and Aripo Rivers are particularly intensively sampled. The Tacarigua, Quare, and Oropouche Rivers are also popular with guppy researchers. Despite their moniker of the millions fish, guppies are not invariably abundant in some of these streams. It can on occasion be difficult to collect the required sample, particularly if individuals of one sex or size class are sought. This raises the spectre of the guppy biologist as an agent of selection alongside the pike cichlid and other predators. Transmission of parasites between rivers and drainages on nets and buckets is another danger. Accidental release of fish into a foreign drainage might also spread disease as well as compromise the genetic distinctness of populations. Although biologists, starting with Caryl Haskins, have manipulated wild populations in various ways, there is no concrete evidence that Trinidadian guppies have been harmed by these activities. Nonetheless, there is a real potential for damage, given the year on year rise in the number of investigators, and increased emphasis on replication across independent units (which is usually taken to mean rivers or drainages). Fig. 7.3 tracks the trend in sampling effort over the last decade and a half. The fact that impacts attributable to scientists are generally minor in comparison to other types of environmental damage does not mean that we should be complacent—nor indeed should we use habitat deterioration as an excuse for irreversible manipulations. It is extremely unlikely that the species will ever be threatened given the guppy's demonstrable ability to colonize and survive in a wide range of freshwater habitats in Trinidad and elsewhere. However, we could unwittingly be compromising the rich variety of populations that attract guppy biologists to Trinidad in the first place. The changes in the genetic characteristics of guppies in one Trinidadian river, wrought by the pioneering transplant experiment, are discussed later in this chapter.

A simple way to monitor the situation would be to develop a system of better, or at least more consistent, reporting of field investigations and collections. Ideally this

Table 7.1 Data reporting of investigations of wild guppy populations

Information to be reported following data gathering
1. Grid reference of site
2. Approximate length (width) of river section examined
3. Basic habitat descriptors, such as temperature and cover
4. Number and sex of fish collected
5. Fate of collected fish, e.g. maintained in captivity, killed for DNA
6. Approximate size (and sex ratio) of population
7. Catch per unit effort (CPUE) statistics
8. Date of study/collection
9. Nature of manipulation, if any
10. Anthropogenic impacts at site, e.g. quarrying, pollution

Notes: Not all categories will be appropriate in all cases, but they could be tailored to individual studies.

should encompass all studies and not just those that make it into the scientific literature. A web-based forum would allow everyone to participate. Such a system would also provide an excellent forum for collaboration. However, published work should also demonstrate good practice, similar to the way in which papers with animal welfare implications are expected to. There are many advantages in having a voluntary scheme devised in cooperation with interested Trinidadian biologists. Not only would this be a real resource in its own right but it might also act as a model for researchers in other systems. It would be a simple means of demonstrating commitment to the conservation of the natural experiment that has already resulted in so many influential papers. Some suggestions for the categories of information that might be usefully recorded are listed in Table 7.1.

7.4 Population viability

Populations of guppies dispersed along linear systems, such as rivers or drainage canals are characterized by considerable rates of gene flow (Haskins *et al.* 1961). Indeed, one of the most intriguing questions regarding the guppy system, and a topic of ongoing research (M. Kinnison, personal communication), is the means by which distinctive traits are maintained in localities, such as those immediately below a barrier waterfall, which receive a rain of immigrants from other populations. Although local extinctions are probably not infrequent events—I have witnessed several that have resulted from drought or pesticide spills—natural levels of dispersal mean that a habitat will often be recolonized soon after it has recovered. Documented introductions confirm that 200 adults can found a viable population. In one case—a garden pond belonging to J. S. Kenny—a thriving colony of guppies descended from a single pregnant female (Carvalho *et al.* 1996). The likelihood of introduced guppies surviving depends on the origin of the fish and the characteristics of the new habitat. For example, a population of 150 guppies obtained from a high-predation locality and introduced into a low-predation one is extremely unlikely to go extinct, whereas the obverse transplant (low-predation founders to high-predation site) has only a 25% chance of surviving for 3 years (Reznick *et al.* 2002*a*, table 2). Guppies are also relatively tolerant of pollution and are one of the last species to persist in seriously impacted sites (Magurran and Phillip 2001*b*). Permanent extinctions are probably rare. However, some populations may be more vulnerable than others, small populations upstream of a barrier in isolated pools or separated by hostile habitat, being obvious examples. A replacement population will not necessarily resemble its predecessor. Local selection acts quickly to mould traits, such as male conspicuousness (Endler 1980) and life-history characteristics (Reznick *et al.* 1990, 1997), but others, such as schooling behaviour (see Fig. 3.4), may be slower to change.

I examined the possible impact of scientific collections by using population viability analysis to model the fate of a pool in the Upper Tunapuna. This stream is typical of sites to which guppy researchers are drawn. It is both unspoilt—indeed beautiful—and accessible. Predation pressure is low; guppies coexist with *Rivulus hartii* and very occasionally with the swamp eel *Synbrachus marmoratus*. The behaviour and ecology of guppies in this stream have been well documented (see, e.g. Magurran and Seghers 1994*c*; Griffiths and Magurran 1997*b*; Evans and Magurran 1999*a*; Godin *et al.* 2003). The pool I have in mind represents the upper natural limit of guppies in this stretch of river. It is separated from lower pools by riffles and small waterfalls, which limit, and possibly entirely prevent, the upstream movement of fish. An investigation using allozymes revealed much lower levels of genetic variation in this pool than in the pools immediately downstream. For example, in contrast to the lower sections of the stream, all guppies in this pool share the same allele at the *Pgm* locus (P. W. Shaw *et al.*, unpublished data), a finding consistent with minimal upstream migration. The pool has an area of 7 m^2, and a maximum depth of 85 cm.

My analysis used VORTEX (Lacy 2000) (www.vortex9.org), one of several approaches to assessing population viability (Brook *et al.* 1999; 2000). Vortex is a stochastic simulation that predicts, among other things, the probability that a population with particular characteristics will go extinct over a specified time period. Although designed for taxa with low fecundity and a long lifespan, the viviparous breeding system of the guppy lends itself to the model. Vortex can be used to model a metapopulation and might therefore be productively applied to guppies occupying a string of pools along a river. As my pool does not receive migrants I treated it as a single population. Vortex takes account of both demographic stochasticity and environmental variability and requires the user to supply information on a range of life history and other variables. Fortunately, good background information on the species in the wild makes it relatively straightforward to estimate the input parameters (see Table 7.2 for details of the values I chose and the information on which I based these choices). Inevitably some of these choices—and particularly the variance element—are educated guesses. I set the 'time period' of the model to be 1 month (instead of the usual year) and ran the simulation for 120 of these intervals, that is 10 years. The probability of population extinction was based on 100 iterations or 'runs'. Extinction

Table 7.2 Values of some of the input parameters used in the population viability analysis of a single, low-predation population of guppies in Trinidad's Northern Range

Input	Value used	Source
Inbreeding depression	Model default	Oosterhout *et al.* 2003*b*
Reproductive system	Polygamous	Houde 1997
Age of first offspring (females)	7 (months)	Reznick *et al.* 1997
Maximum age of reproduction	24 (months)	Data for this population
Maximum number of progeny (per month)	15	Data for this population
Sex ratio at birth (% males)	50	Pettersson *et al.* 2004
Offspring per female (±s.d.)	5.56±1.73	Data for this population
% Adult females breeding	90%	Becher and Magurran 2004
Mortality of juvenile fish	85%	Reznick *et al.* 1996*b*
Female (adult) mortality	10%	Reznick *et al.* 1996*b*
Male (adult) mortality	15%	Reznick *et al.* 1996*b*
Catastrophe	Severe flood once in 15 years (removes over 75% of fish)	Data for this population
Initial population size	200	Arbitrary value
Carrying capacity	600	Data for this population

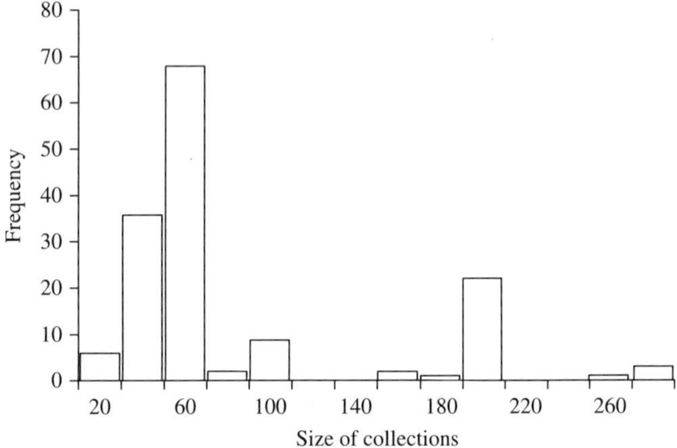

Fig. 7.4 Frequency distribution of the size (number of individuals) of reported guppy collections.

was assumed to take place if the population diminished to fewer than five individuals. Finally, I examined the consequences of scientific sampling by assuming that 20 adult males and 20 adult females were removed (harvested) from the population once every 12 months. This is towards the lower end of the sample sizes collected by scientists in Trinidad (Fig. 7.4).

The simulations showed a slight, but potentially important, increase in extinction probability when small numbers of fish were sampled on an annual basis. The median probability of extinction (PE) in the absence of harvesting was 0.03; this rose to 0.09

when collecting occurred. In other words, a 3% chance of a natural extinction in 10-year period increases to a 9% chance if fish are systematically removed. This is an illustrative example rather than a formal analysis but it does demonstrate how population viability analysis can be applied to the species. (Another interesting exercise is to examine the viability of populations founded by different numbers of individuals.) The outcome depends on how the population is characterized; changes in variables, such as carrying capacity, change the likelihood of extinction. I have deliberately chosen a simple (but genuine) population, as well as one that, because of its isolation, is particularly vulnerable to extinction. Nonetheless, the results demonstrate that investigators have the potential to impact the populations they come to study. Repeated and uncoordinated, collecting by different groups at the same site could amplify the problem. There is evidence that researchers are cognizant of the dangers of overcollecting. For example, O'Steen *et al.* (2002) report that in 1997, population sizes in the Yarra River were so small as to cause them to modify their protocol. A different type of researcher impact is reviewed in the next section.

7.5 Long-term consequences of artificial introductions

Transplant experiments are a powerful means of investigating evolution in the wild. However, artificial introductions of guppies leave an indelible mark on the manipulated assemblages. The enduring consequences of these experiments are particularly stark in the Turure River. As noted in Chapters 2 and 3, Caryl Haskins moved 200 guppies from a high-predation section of the Arima River in the Caroni Drainage to a guppy-free stretch of the Upper Turure in the Oropouche Drainage (Fig. 7.5). The descendants of the transplanted fish can still be found in the Upper Turure (points T1 and T2 in Fig. 7.5). (Magurran *et al.* 1992; Shaw *et al.* 1992; Becher and Magurran 2000; Russell 2004). However, the Caroni genes that Haskins introduced have not been confined to the upstream portion of the Turure River. There is marked asymmetrical introgression by introduced nuclear alleles ($\geq$90%) in the downstream stretch of the river (T3 and T4 in Fig. 7.5, Russell 2004). This suggests that the native nuclear genotype below the barrier waterfall has been entirely eradicated. The introduced mitochondrial haplotype, on the other hand, declines from 84% below the barrier waterfall (T3) to 42% near the confluence with the Quare (T4). There are several explanations for the discrepant patterns in mitochondrial and nuclear DNA. These include stochastic effects, male-biased dispersal, and enhanced mating success by migrant males (Russell 2004). Support for the latter two hypotheses comes from the observation that males are more mobile than females (Magurran and Seghers 1994*c*; Croft *et al.* 2003*b*) and that Caroni males are more successful than Oropouche males during mating competition (Magurran *et al.* 1996). It appears that Caroni genes are continuing to infiltrate the Oropouche system. Shaw *et al.* (1992) detected no diagnostic Caroni alleles at T4. Shaw *et al.* used allozymes, which are not directly comparable with the micosatellite markers employed by Russel. Nonetheless, the outcome is consistent with substantial gene flow in the last decade. Although dispersal is predominantly downstream, gene flow may occur upstream in the absence of

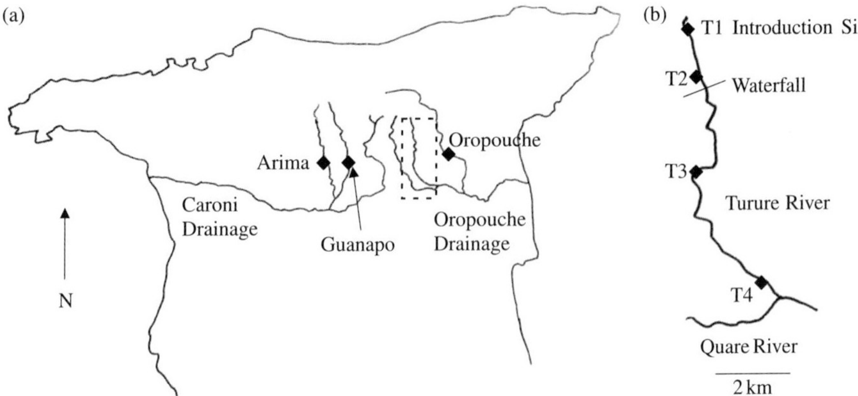

Fig. 7.5 Caryl Haskins's transplant experiment. (a) Location of key populations in the Caroni and Oropouche drainages. The Turure River, identified by the dashed box, is shown in detail. In (b) 200 guppies were collected from the Lower Guanapo (not the Lower Arima as originally thought) in 1957 and introduced to the upper section of the Turure River (T1). Fish above the barrier waterfall (T1 and T2) are the direct descendants of these transplanted fish. Guppies below the barrier waterfall are an admixture of Caroni and Oropouche genotypes. See text for details. After Russell (2004).

barriers (Haskins *et al.* 1961). We might expect, therefore, that the genes introduced by Haskins in 1957 will gradually invade the lower Quare and Lower Oropouche Rivers.

It is arguable that guppy transplants affect no one other than the biologists who come to study the species. It seems unlikely, for example, that assemblage structure, or productivity, or the behaviour of predators and competitors, will be changed as a result of transplants, such as the Haskins one. But the fact that the character of the most divergent clade of guppies is being slowly modified as the result of a small transplant that took place almost half a century ago is cause for reflection. The consequences of within-drainage or within-river transplants will not be as dramatic of course, but they will have an impact. And, while fully acknowledging that our understanding of evolution in the wild would have been impeded, were it not for the Haskins transplant and its successors, I propose that no further transplants should be undertaken without consultation with the scientific community and the Trinidadian authorities. Indeed, it might be best to continue to investigate existing transplants but to desist from new ones and to switch to other methods, such as mesocosm or greenhouse experiments, instead.

7.6 Exotic guppies

Thus far I have emphasized the risks facing guppy populations within their natural range. It would, however, be remiss to conclude without pointing out that guppies can themselves participate in Diamond's evil quartet in the guise of an invasive species. The literature is scant, and the area ripe for research. Accumulating data nonetheless, suggest that guppies adversely affect other species. The Goodeinae, a group of viviparous fish endemic to Central Mexico (Webb *et al.* 2004), provide one example. Several goodeid species are confined to single localities, and 13 out of the 35 species in the family are regarded as endangered. One species exists only in zoo collections in the USA and Europe. The threats facing the Goodeinae include habitat fragmentation, habitat deterioration, and the introduction of exotic fish species (De La Vega-Salazar *et al.* 2003*a, b*). Guppies have been introduced either deliberately (for mosquito control) or accidentally (as a by-product of the aquarium trade) into a number of localities where vulnerable goodeids are found. For instance, guppies are present in La Mintzita spring reservoir, one of the few remaining habitats of *Skiffia lermae*, and in the only remaining locality of *Zoogoneticus tequila* in the wild (C. Macías-Garcia, personal communication). Other instances of likely problems include Hong Kong, where the native minnow *Aphyocypris lini* appears seriously threatened by introduced guppies (Man and Hodgkiss 1981), Papua New Guinea (Allen 1991), and the Philippines (Juliano *et al.* 1989). Reports of guppy introductions often mention possible ecological damage (Welcomme 1988). The guppy has been introduced to more places around the world than any other poeciliid with the exception of the mosquito fish (*Gambusia affinis*) (Courtenay and Meffe 1989).

The manner in which invasive fish species impact native assemblages is incompletely understood. Much of the headline concern about invasives relates to predation—the decline of endemic cichlids in Lake Victoria and other African lakes as a result of the introduction of the Nile Perch, *Lates niloticus*, being a particularly well-cited example (Ogutu-Ohwayo 1990, 1993). Smaller-bodied invasives, including most poeciliids, are unlikely to be serious predators—though they may eat eggs and small fry. Instead, competition for food or space, and disruption of activities, such as courtship, probably account for their negative impact. This may be compounded by anthropogenic changes to the habitat, such as a rise in water temperature, that favour the invasive at the expense of the native. Even small reductions in the population size or changes in population structure, of indigenous fish could increase their risk of extinction.

7.7 Conclusions

Guppies are not an endangered species, nor are they ever likely to become one. But there are still grounds for concern about their future. This book has, I hope, illustrated the influential role that guppies have played in shaping evolutionary biology. This influence derives largely from the fact that it has been possible to interpret findings in the context of the natural habitat and ecological communities in which these fish live. For historical reasons much of this work has been focussed on populations native to a geographically restricted area of Trinidad. Unfortunately the Northern Range in Trinidad, to which most researchers are drawn, is vulnerable to habitat deterioration and habitat loss. As guppies are the most tenacious of freshwater species in Trinidad, fish in most localities will probably be able to resist extinction. What is at risk is the ecological tapestry in which these populations are embedded. The danger is not so much one of extermination as it is of homogenization. Researchers potentially exacerbate these problems by over-collecting and by fish movements. Since interest in Trinidadian guppies shows no sign of abating I propose that we as scientists become proactive in safeguarding the system that is so rewarding to study.

The broader consequences of introductions are clear from the impact that guppies can have when they become established in non-native habitats. One problem here has been the misguided assumption that because the fish are small they pose no threat. It is likely that their role as invasives will soon attract considerable attention. Baseline information on guppy behaviour and ecology will be an invaluable asset in the endeavour. At the same time guppies have an, as yet little exploited, potential to contribute to the science of conservation biology. This chapter has highlighted some opportunities. The emerging discipline of conservation behaviour (Blumstein and Fernández-Juricic 2004) is one for which guppies are particularly well suited. Guppies then, occupy the triple roles of a species of conservation concern, a taxon that can adversely affect other species, and a model for conservation research. In this they illustrate well the diversity of issues that confront conservation biologists.

8

Postscript and prospects

'However, to be frank, it has yet to be shown that guppies do anything interesting' asserted McGrew (2004) in his review of the book *Animal innovations* (Reader and Laland 2003). While I concede that guppy innovation and intelligence is not on par with that exhibited by chimps, or possibly even canaries, this book has attempted to highlight the rich diversity of behaviour that guppies do display. Research has shown that guppies recognize individuals and make strategic decisions based on the past behaviour of familiar conspecifics. These small fish are also capable of learning quite complex tasks and making subtle discriminations. This behaviour is played out against a backdrop of population differentiation that makes the system ideally suited for investigations of natural and sexual selection. It is no accident that pioneering research on topics as disparate as sperm competition and female choice was inspired by this species. In this final chapter, I attempt to draw together some of the diverse themes in the book. I first revisit my initial premise—that good data are durable, and that information collected using straightforward, often very low-tech, approaches have stood the test of time. I make the case that knowing the organism, and the insights that flow from a detailed understanding of its behaviour and ecology, underpin significant advances in knowledge. This is not to say that new techniques are unimportant. In fact the guppy illustrates well the ability of modern technologies, such as microsatellite markers, to answer previously intractable questions. Equally, new developments in theory have stimulated innovative research. But my point is that these recent developments are powerful precisely because they are explored in the context of a well-characterized landscape. Next I consider some of the issues raised by the early investigators and ask whether they have been resolved. From this flows a series of unresolved questions. I attempt to identify some potentially productive lines of research, bearing in mind the need to protect the system. I also reflect on experimental design and consider why, seemingly similar investigations can sometimes produce perplexingly different outcomes. Finally, I emphasize how the separate strands of research on guppies add up to more than the sum of the parts.

8.1 The test of time

One of the most fascinating aspects of the large literature on guppies is the way in which it reflects changing fashions in evolutionary ecology. Indeed, many of the early conclusions seem dated and even misguided. Schmidt (1919*a*, *b*, 1920), for example, was concerned with racial differences and concluded:

> this brings us then to the apparently remarkable state of things, that racial differences here exist only in the one sex, the male, no such idea being tenable in the case of the opposite sex
> (Schmidt 1920, p. 11)

Setting aside the question of whether race is a concept that is meaningfully applied to guppy populations, it is clear to us now that male colour patterns are the product of female choice and that females carry some of the genes that maintain colour patterns in a population. However, Schmidt's interpretation of his results, which was shaped by prevailing ideas about genetics, does not detract from the many other insights he gleaned from his data. His observations on sperm competition and Y-linked inheritance were remarkable and have stood the test of time. Likewise, the Haskinses (Haskins and Haskins 1949) initially concluded that it was males rather than females that did most of the choosing. However, the same paper led directly to the research that revealed the significance of female choice. It further established the capacity of males for learned mate discrimination. Early research papers, such as these, are thus much more than a historical record as they contain many robust conclusions and include thoughtful, but often overlooked, discussions of topics, such as mating strategy and population biology. This literature can also be a fertile source of ideas for new investigations as well as be a repository of unique datasets.

8.2 Future directions

Despite intensive research, there are many questions that remain unanswered. The next section reviews some of the directions that research into the evolutionary ecology of guppies might take in the future.

8.2.1 The guppy genome

Fish provide some of the most successful genomics models. However, notwithstanding the wealth of information on the genomes of species, such as zebrafish, *Danio rerio*, pufferfish, *Fugu rubripes* and medaka, and *Oryzias latipes*, very little is known about the genetic basis of ecological or evolutionarily significant behaviours in fish. There are but a handful of studies on aggression and shoaling in zebrafish (Pritchard *et al.* 2001; Gerlai 2003; Wright *et al.* 2003)—a species that seems ideal for this type of research. A historical bias towards developmental research is one reason for this omission, the other is the paucity of information on the behaviour of these species in the wild, together with a lack of data on natural variation that could be exploited by a functional genomics approach. The ecological conditions experienced by the populations from which stocks of zebrafish are sourced are generally unknown. It seems probable, based on ease of access, that most lines are descended from fish collected in paddy fields. The uniform nature of these habitats will dampen variation among populations, even those that are geographically dispersed. Traits, such as anti-predator behaviour, tend to diminish after several years in captivity in any case (Kelley and Magurran 2003*a*), and the tightly controlled and cramped environments in which these fish are usually reared will also select against natural variation in behaviour. It is unlikely that any significant advances in elucidating the genetic architecture of zebrafish behaviour can be made until we learn more about its evolutionary ecology in the wild. We have the opposite problem where guppies are concerned. Guppy research, while largely predicated on adaptive variation among populations, is disadvantaged by the fact that virtually nothing is known about the genes that underpin these traits. Fortunately, work to establish a genetic linkage map based on single nucleotide polymorphisms (SNPs) in expressed genes is now underway.[3] This resource will finally allow researchers to identify the genes responsible for phenotypic variation.

8.2.2 *Success of sneaky mating*

As Chapter 4 pointed out, the success of sneaky mating relative to consensual copulation remains unclear despite a growing literature showing that the relative frequency with which this tactic is employed varies adaptively in relation to predation risk and to the opportunity for sexual selection (e.g. Magurran and Seghers 1990*c*; Jirotkul 2000*a*, *b*). Although there are considerable technical challenges to be overcome, a comparative analysis, across populations and poeciliid species, of the contribution that sneaking makes to paternity would be invaluable.

8.2.3 *Dynamics of sperm competition*

Investigation of sperm competition in guppies has a long history (Schmidt 1920) and the species continues to provide seminal insights into the mechanisms involved. Artificial insemination has proved a powerful tool here, particularly through its ability to remove female choice and mating order effects (e.g. Evans *et al.* 2003*b*). Histological techniques have also been used to considerable effect (Kobayashi and Iwamatsu 2002). But there is still much to be learnt. For example, we know relatively little about how the new and old sperm interact, or how post-copulatory choice is exerted. Evidence that gametic incompatibility arises when populations have spent extended periods in allopatry (Chapter 6) shows that subtle interactions between eggs and sperm are possible, though the actual mechanisms involved remain unknown. One aspect of guppy (indeed poeciliid) reproductive biology that may be pertinent here is the insemination of sperm bundles rather than single sperm. Do these bundles have a mechanical role in sperm competition? Although they appear to break down quickly in the female gonoduct, bundles may impede rival sperm. What happens when sperm reach the ovaries and how exactly are the sperm storage sites populated? Why is it that older, stored, sperm, are less competitive than fresh inseminates? And

[3] http://www.weigelworld.org/research/projects/guppyvariation/

why do sperm delivered through consensual copulations appear to be more successful than those that result from sneaky matings? The biochemical and immunological interactions between rival sperm, and between sperm and egg, also remain to be resolved. MHC-mediated interactions may well prove crucial here.

8.2.4 *Maternal investment versus genetic sire effects*

Accumulating results show that females that mate with preferred males, and engage in polyandry, produce superior offspring (Evans and Magurran 2000; Evans *et al.* 2004*b*; Ojanguren *et al.* 2005). It is uncertain, however, whether these findings can be explained by differential allocation (Sheldon 2000), whereby the investment a female makes in her offspring is mediated by her perception of the attractiveness of her partners, or by the superior genetic contribution of preferred males, or indeed by a combination of both. Research has shown that female poeciliids have some flexibility regarding offspring provisioning (Reznick and Yang 1993; Trexler 1997; Trexler and DeAngelis 2003) and may be able to delay fertilization (Evans and Magurran 2000). These issues, like those raised above, will require a much more detailed understanding of post-copulatory events.

8.2.5 *Ontogeny*

In contrast to the zebrafish, on which developmental research is conducted on an industrial scale, very little is known about the early life stages and ontogenetic changes in guppies. One practical reason for this is internal fertilization, which makes it difficult to directly observe development, or to manipulate embryos. Fortunately, *in vitro* rearing techniques are now being developed (C. Dreyer, personal communication). It would be interesting to determine the consequences of varying gestation times and differences in offspring size at birth, produced by selection on life history, in relation to behavioural traits, including shoaling, and anti-predator responses, and ecological interactions, such as competition. In addition there have been relatively few investigations of how early experience impacts adult behaviour (see Liley 1966; Breden *et al.* 1995; Rosenqvist and Houde 1997 for some exceptions) though data on other fish species show that these can be profound. Guppies would also be an ideal taxon in which to investigate the extent to which behavioural syndromes—suites of correlated behaviours across contexts (Sih *et al.* 2004, p. 372)—shape individual behaviour. It is already known that aggression and schooling behaviour covary across populations (Magurran and Seghers 1991) but the genetic and developmental basis of this correlation have not yet been explored.

8.2.6 *Lifetime reproductive success*

Laboratory studies (Becher and Magurran 2004; Reznick *et al.* 2004) are beginning to provide some indications of patterns of lifetime reproductive success in both males and females. Recent advances in molecular techniques and the availability of increasingly numbers of microsatellite markers mean that it is increasingly feasible to investigate lifetime reproductive success in the wild or in near natural environments.

8.2.7 Predation risk

As Chapter 2 indicated, variation in predation risk is the factor that underpins much of the research on guppies in Trinidad, yet, the nature of that risk is very incompletely understood. It would be extremely interesting to determine how risk varies over time as well as over space, the contributions of various aerial and aquatic predators, and the degree of risk experienced at different life stages.

8.2.8 Multiple cues in predator evasion and mate choice

With a few notable exceptions, investigations of predator assessment tend to focus on either olfactory or visual cues. The same can be said of mate choice studies. Consideration of multiple cues, including tactile ones, is warranted. Since some guppy populations inhabit naturally turbid waters it would be interesting to determine how visual, olfactory, and tactile information are ranked in different environments, and whether the same rankings are used in both predator detection and mating decisions.

8.3 Some thoughts on experimental design

Although it is something I had been aware of, it was only when revisiting the literature during the writing of this book that I became fully conscious of how small differences in experimental design can have substantial implications for the outcome of an investigation. Indeed the guppy literature provides multiple instances of contrasting conclusions derived from experiments that appear, on the surface at least, to be nearly identical. For instance, Dill *et al.* (1999) and Evans *et al.* (2002*a*) recorded different responses of males to females that had detected predators (see Chapter 4). Small variations in experimental design and in the prior experience of the participating fish, could account for this. Intrinsic population differences—even within equivalent predation regimes—may also be held responsible. Another example concerns the change in female choice behaviour in the presence of predators (Godin and Briggs 1996; Gong and Gibson 1996; and see Chapter 4). The manner in which a predation threat is presented appears to be very important here. Fish rapidly habituate to dummies, particularly unrealistic ones (Magurran and Girling 1986), may respond inappropriately if certain cues are missing, and probably rapidly learn that a live predator behind glass does not present a serious threat. This raises two issues. First, we have no standard method of calibrating predation risk. Most people adopt the pragmatic approach of doing something that works, and choose a form of threat that frightens the fish enough to change their behaviour significantly, in the context of a particular

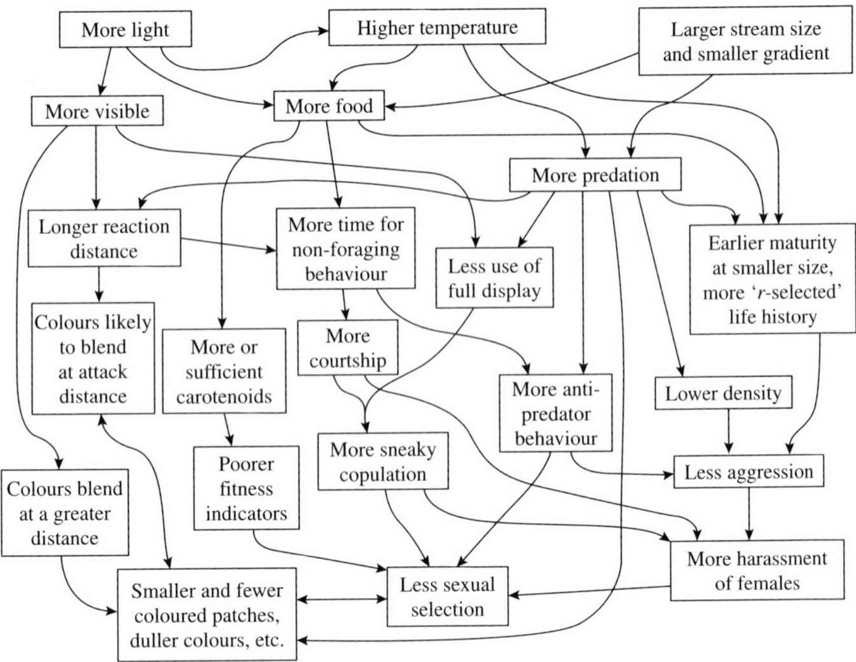

Fig. 8.1 Multiple ways in which environmental factors influence a variety of guppy traits in a typical high locality in Trinidad. After figure 1 in Endler (1995). Reproduced with kind permission of John Endler, and of Elsevier.

experiment. However, it is very difficult to ensure equivalence among investigators and this can hinder comparative work. A related issue is the use of live predators, something that increasing numbers of researchers shun for animal welfare and/or for legal reasons. Again, the size, behaviour, experience of the predators, and type of arena in which fish are tested will influence the results. These variables probably account for the different conclusions in the literature regarding size selectivity by predators and relative vulnerability of the sexes. Second, the multiplicity of responses by guppies to variable predation risk in the laboratory is almost certainly apparent in the wild too. Thus, it may be possible to make a virtue out of the different ways in which captive fish respond to predation risk and use this to understand more fully the nature of anti-predator responses in the wild. Similar observations can be made of other types of behavioural work, of course. For example, prior experience influences mating preferences by females (Breden *et al.* 1995; Rosenqvist and Houde 1997), choices may be reversed following social interactions (Dugatkin and Godin 1992*c*), and simultaneous versus sequential presentation of males can result in different decisions. Standard methods for the measurement of choice are increasingly being adopted (Houde 1997; Brooks and Endler 2001*b*).

8.4 More than the sum of the parts

It is not only animals that make trade-offs. Scientists too are limited by time and by what is practical. Guppy researchers, like those investigating other single species systems, sacrifice inter-taxa comparisons, except through the literature, for a more detailed understanding of the group. But the flip side of this is that it becomes possible to make links across the different domains of evolutionary ecology and to ask, for example, how anti-predator responses mediate mate choice and how life-history decisions in turn moderate behaviour. This fine level of resolution prompted John Endler (1995) to examine the co-evolution of multiple traits and to trace thc manifold consequences of environmental variation. Fig. 8.1 illustrates how his exploration of a network of relationships can lead to a much deeper understanding of functional and evolutionary interactions. This type of analysis is also invaluable in generating novel hypotheses. I am confident that the guppy system will continue to offer unrivalled opportunities to test theories in evolutionary ecology and that it will generate new data and insights that have relevance in a broad range of fields. The real challenge will be to ensure that this unique natural laboratory is preserved.

References

Abrahams, M. V. 1989. Foraging guppies and the ideal free distribution: the influence of information on patch choice. *Ethology* **82**, 116–126.

Abrahams, M. V. 1993. The trade-off between foraging and courting in male guppies. *Animal Behaviour* **45**, 673–681.

Abrahams, M. V. and Dill, L. M. 1989. A determination of the energetic equivalence of the risk of predation. *Ecology* **70**, 999–1007.

Abrams, P. 1993. Does increased mortality favor the evolution of more rapid senescence? *Evolution* **47**, 877–887.

Abrams, P. A. 2004. Mortality and lifespan. *Nature* **431**, 1048–1049.

Abrams, P. A. and Rowe, L. 1996. The effects of predation on the age and size of maturity of prey. *Evolution* **50**, 1052–1061.

Albers, P. C. H. 2000. Evidence of evolution of guppies in a semi-natural environment. *Netherlands Journal of Zoology* **50**, 425–433.

Alexander, H. J. and Breden, F. 2004. Sexual isolation and extreme morphological divergence in the Cumaná guppy: a possible case of incipient speciation. *Journal of Evolutionary Biology* **17**, 1238–1254.

Alkins-Koo, M. 2000. Reproductive timing of fishes in a tropical intermittent stream. *Environmental Biology of Fishes* **57**, 49–66.

Allen, G. R. 1991. *Field guide to the freshwater fishes of New Guinea*. Madang, Papua New Guinea: Christensen Research Institute.

Amundsen, T. 2003. Fishes as models in studies of sexual selection and parental care. *Journal of Fish Biology* **63** (Suppl. A), 17–52.

Andersson, M. 1994. *Sexual selection*. Princeton, New Jersey: Princeton University Press.

Andreev, O. A. 1994. The behavior of male and female guppies, *Poecilia reticulata*, in an unfamiliar environment. *Journal of Icthyology* **34**, 139–143.

Anstis, S., Hutahajan, P., and Cavanagh, P. 1998. Optomotor test for wavelength sensitivity in the guppyfish (*Poecilia reticulata*). *Vision Research* **38**, 45–53.

Archer, S. N. and Hirano, J. 1997. Opsin sequences of rod visual pigments in two species of Poeciliid fish. *Journal of Fish Biology* **51**, 215–217.

Arendt, J. D. and Reznick, D. N. 2005. Evolution of juvenile growth rate in female guppies (*Poecilia reticulata*): predator regime or resource level? *Proceedings of the Royal Society of London, Series B* **272**, 333–337.

Avise, J. C. 1994. *Molecular markers, natural history, and evolution*. New York: Chapman and Hall.

Avise, J. C. and Shapiro, D. Y. 1986. Evaluating kinship of newly settled juveniles within social groups of the coral reef fish *Anthias squamipinnis*. *Evolution* **40**, 1051–1059.

Baatrup, E. and Junge, M. 2001. Antiandrogenic pesticides disrupt sexual characteristics in the adult male guppy (*Poecilia reticulata*). *Environmental Health Perspectives* **109**, 1063–1070.

Baerends, G. B., Brouwer, R., and Waterbolk, H. T. 1955. Ethological studies in *Lebistes reticulatus* (Peters). I. An analysis of the male courtship pattern. *Behaviour* **8**, 249–335.

Bakker, T. C. M. and Pomiankowski, A. 1995. The genetic basis of female mate preferences. *Journal of Evolutionary Biology* **8**, 129–171.

Barlow, J. 1992. Nonlinear and logistic growth in experimental populations of guppies. *Ecology* **73**, 941–950.

Barus, V., Penaz, M., and Wohlgemuth, E. 1995. Morphometry of Cuban population of the guppy, *Poecilia reticulata* (Pisecs: Poeciliidae). *Folia Zoologica* **44**, 363–370.

Barus, V. and Wohlgemuth, E. 1993. Study on the collection of freshwater fishes from eastern Cuba with taxonomical nores. *Folia Zoologica* **42**, 63.

Basola, A. L. 1990. Female preference predates the evolution of the sword in swordtail fish. *Science* **250**, 808–810.

Basolo, A. L. 1995. Phylogenetic evidence for the role of a pre-existing bias in sexual selection. *Proceedings of the Royal Society of London, Series B* **259**, 307–311.

Bateman, A. J. 1948. Intra-sexual selection in *Drosophila*. *Heredity* **2**, 349–368.

Bates, L. and Chappell, J. 2002. Inhibition of optimal behavior by social transmission in the guppy depends on shoaling. *Behavioral Ecology* **13**, 827–831.

Bayley, M., Nielsen, J. R., and Baatrup, E. 1999. Guppy sexual behavior as an effect biomarker of estrogen mimics. *Ecotoxicology and Environmental Safety* **43**, 68–73.

Bayley, M., Jumge, M., and Baartrup, E. 2002. Exposure of juvenile guppies to three antiandrogens causes demasculinization and a reduced sperm count in adult males. *Aquatic Toxicology* **56**, 227–237.

Bayley, M., Larsen, P. F., Baekgaard, H., and Baatrup, E. 2003. The effects of vinclozlin, an anti-androgenic fungicide, in male guppy secondary sex characters and reproductive success. *Biology of Reproduction* **69**, 1951–1956.

Bayley, P. B. and Li, H. W. 1992. Riverine fishes. In *The rivers handbook* (ed. P. Calow and G. E. Petts), pp. 251–281. Oxford: Blackwell Science.

Becher, S. A. and Magurran, A. E. 2000. Gene flow in Trinidadian guppies. *Journal of Fish Biology* **56**, 241–249.

Becher, S. A. and Magurran, A. E. 2004. Multiple mating and reproductive skew in Trinidadian guppies. *Proceedings of the Royal Society of London, Series B* **271**, 1009–1014.

Becher, S. A., Russell, S. T., and Magurran, A. E. 2002. Isolation and characterisation of polymorphic microsatellites in the Trinidadian guppy (*Poecilia reticulata*). *Molecular Ecology Notes* **2**, 456–458.

Benz, J. J. and Leger, D. W. 1992. Evidence for mate choice by male guppies (*Poecilia reticulata*). *Transactions of the Nebraska Academy of Sciences* **20**, 47–51.

Berkeley, S. A., Chapman, C., and Sogard, S. M. 2004. Maternal age as a determinant of larval growth and survival in a marine fish, *Sebastes melanops*. *Ecology* **85**, 1258–1264.

Bertalanffy, L. v. 1938. A quantitative theory of organic growth (inquiries on growth laws, II). *Human Biology* **10**, 181–213.

Bildsoe, M. and Sorensen, J. B. 1994. A method of modelling time-dependent data—swimming in guppies (*Poecilia reticulata*) under threat of a predator. *Behavioural Processes* **31**, 75–96.

Billard, R. 1969 La sperm spermatogenese de *Poecilia reticulata*. *Annales de Biologie animale Biochimie Biophysique* **9**, 251–271.

Billard, R. 1986. Spermatogenesis and spermatology of some teleost fish species. *Reproduction, Nutrition, Développment* **26**, 877–920.

Birkhead, T. 2000*a*. *Promiscuity*. London: Faber and Faber.

Birkhead, T. R. 2000*b*. Defining and demonstrating postcopulatory female choice—again. *Evolution* **54**, 1057–1060.

Birkhead, T. R. and Pizzari, T. 2002. Postcopulatory sexual selection. *Nature Review Genetics* **3**, 262–273.

Bisazza, A. 1993. Male competition, female mate choice and sexual size dimorphism in poeciliid fishes. In *Behavioural ecology of fishes* (ed. F. A. Huntingford and P. Totticelli), pp. 257–286. Chur, Switzerland: Harwood Academic.

Bisazza, A. and Pilastro, A. 1997. Small size mating advantage and reversed size dimorphism in poeciliid fishes. *Journal of Fish Biology* **50**, 397–406.

Bisazza, A., Pignatti, R., and Vallortigara, G. 1997. Laterality in detour behaviour: interspecific variation in poeciliid fish. *Animal Behaviour* **54**, 1273–1281.

Bischof, H.-J. 1994. Sexual imprinting as a two-stage process. In *Causal mechanisms of behavioural development* (ed. J. A. Hogan and J. J. Bolhuis), pp. 82–97. Cambridge: Cambridge University Press.

Bischoff, R., Gould, J. L., and Rubenstein, D. I. 1985. Tail size and female choice in the guppy (*Poecilia reticulata*). *Behavioral Ecology and Sociobiology* **17**, 253–255.

Blows, M. W., Brooks, R., and Kraft, P. G. 2003. Exploring complex fitness surfaces: multiple ornamentation and polymorphism in male guppies. *Evolution* **57**, 1622–1630.

Blumstein, D. T. and Fernández-Juricic, E. 2004. The emergence of conservation behavior. *Conservation Biology* **18**, 1175–1177.

Boos, H. E. A. 1984. A consideration of the terrestrial reptile fauna on some offshore islands north west of Trinidad. *Living World* **1983/1984**, 19–26.

Bozynski, C. C. and Liley, N. R. 2003. The effect of male presence on spermiation, and of male sexual activity on 'ready' sperm in the male guppy. *Animal Behaviour* **65**, 53–58.

Braun, K. and Harper, R. G. 1993. Effects of predator size and female receptivity on courtship behavior of captive bred male guppies. *Transactions of the Illinois State Academy of Science* **86**, 127–132.

Breden, F. and Bertrand, M. 1999. A test for female attraction to male orange coloration in *Poecilia picta*. *Environmental Biology of Fishes* **55**, 449–453.

Breden, F. and Hornaday, K. 1994. Test of indirect model of selection in the Trinidad guppy. *Heredity* **73**, 291–297.

Breden, F. and Stoner, G. 1987. Male predation risk determines female preference in the guppy. *Nature* **329**, 831–833.

Breden, F., Scott, M. A., and Michel, E. 1987. Genetic differentiation for antipredator behaviour in the Trinidad guppy, *Poecilia reticulata*. *Animal Behaviour* **35**, 618–620.

Breden, F., Novinger, D., and Schubert, A. 1995. The effect of experience on mate choice in the Trinidad guppy, *Poecilia reticulata*. *Environmental Biology of Fishes* **7**, 323–328.

Breden, F., Ptacek, M. B., Rashed, M., Taphorn, D., and Figueiredo, C. A. 1999. Molecular phylogeny of the live-bearing fish genus *Poecilia* (Cyprinidontiformes: Poeciliidae). *Molecular Phylogenetics and Evolution* **12**, 95–104.

Breder, C. M. 1951. Studies on the structure of fish shoals. *Bulletin of the American Museum of Natural History* **98**, 1–27.

Breder, C. M. and Coates, C. W. 1935. Sex recognition in the guppy, *Lebistes reticulatus* Peters. *Zoologica* **19**, 187–207.

Brewster, J. and Houde, A. 2003. Are female guppies more likely to flee when approached by two males? *Journal of Fish Biology* **63**, 1056–1059.

Briggs, S. E., Godin, J.-G. J., and Dugatkin, L. A. 1996. Mate-choice copying under predation risk in the Trinidadian guppy (*Poecilia reticulata*). *Behavioral Ecology* **7**, 151–157.

Bronikowski, A. M., Clark, M. E., Rodd, F. H., and Reznick, D. N. 2002. Population-dynamic consequencs of predator-induced life history variation in the guppy (*Poecilia reticulata*). *Ecology* **83**, 2194–2204.

Brook, B. W., Cannon, J. R., Lacy, R. C., Mirande, C., and Frankham, R. 1999. Comparison of the population viability analysis packages GAPPS, INMAT, RAMAS and VORTEX for the whooping crane (*Grus americana*). *Animal Conservation* **2**, 23–32.

Brook, B. W., O'Grady, J. J., Chapman, A. P., Burgman, M. A., Akcakaya, H. R., and Frankham, R. 2000. Predictive accuracy of population viability analysis in conservation biology. *Nature* **404,** 385–387.

Brooke, A. P. 1994. Diet of the fishing bat, *Noctilio leporinus* (Chiroptera, Noctilionidae). *Journal of Mammalogy* **75**, 212–218.

Brooke, A. P. 1997. Organization and foraging behaviour of the fishing bat, *Noctilio leporinus* (Chiroptera: Noctilionidae). *Ethology* **103**, 421–436.

Brooks, R. 1996. Copying and the repeatability of mate choice. *Behavioral Ecology and Sociobiology* **39**, 323–329.

Brooks, R. 1998. The importance of male copying and cultural inheritance of mating preferences. *Trends in Ecology and Evolution* **13**, 45–46.

Brooks, R. 2000. Negative genetic correlation between male sexual attractiveness and survival. *Nature* **406**, 67–70.

Brooks, R. 2002. Variation in female mate choice within guppy populations: population divergence, multiple ornaments and the maintenance of polymorphism. *Genetica* **116**, 343–358.

Brooks, R. and Caithness, N. 1995*a*. Female choice in a feral guppy population: are there multiple cues? *Animal Behaviour* **50**, 301–307.

Brooks, R. and Caithness, N. 1995*b*. Female guppies use orange as a mate choice cue: a manipulative test. *South African Journal of Zoology* **30**, 200–201.

Brooks, R. and Caithness, N. 1995*c*. Manipulating a seemingly nonpreferred male ornament reveals a role in female choice. *Proceedings of the Royal Society of London, Series B* **261**, 7–10.

Brooks, R. and Endler, J. A. 2001*a*. Direct and indirect selection and quantitative genetics of male traits in guppies (*Poecilia reticulata*). *Evolution* **55**, 1002–1015.

Brooks, R. and Endler, J. A. 2001*b*. Female guppies agree to differ: phenotypic and genetic variation in mate-choice behavior and the consequences for sexual selection. *Evolution* **55**, 1644–1655.

Brosnan, S. F., Earley, R. L., and Dugatkin, L. A. 2003. Observational learning and predator inspection in guppies (*Poecilia reticulata*). *Ethology* **109**, 823–833.

Brown, C. and Laland, K. N. 2002. Social learning of a novel avoidance task in the guppy: conformity and social release. *Animal Behaviour* **64**, 41–47.

Brown, C. and Laland, K. N. 2003. Social learning in fishes: a review. *Fish and Fisheries* **4**, 280–288.

Brown, G. E. 2003. Learning about danger: chemical cues and local risk assessment in prey fishes. *Fish and Fisheries* **4**, 227–234.

Brown, G. E. and Godin, J.-G. J. 1999. Chemical alarm signals in wild Trinidadian guppies (*Poecilia reticulata*). *Canadian Journal of Zoology* **77**, 562–570.

Brown, L. P. 1982. Can guppies adjust the sex ratio? *American Naturalist* **120**, 694–698.

Bruce, K. E. and White, W. G. 1995. Agonistic relationships and sexual behaviour patterns in male guppies, *Poecilia reticulata*. *Animal Behaviour* **50**, 1009–1021.

Bryant, M. J. and Reznick, D. N. 2004. Comparative studies of senescence in natural populations of guppies. *American Naturalist* **163**, 55–68.

Brönmark, C. and Hansson, L. A. 2000. Chemical communication in aquatic systems: an introduction. *Oikos* **88**, 103–110.

Brönmark, C. and Miner, J. G. 1992. Predator-induced phenotypical change in body morphology in crucian carp. *Science* **258**, 1348–1350.

Brönmark, C. and Pettersson, L. B. 1994. Chemical cues from piscivores induce a change in morphology in crucian carp. *Oikos* **70**, 396–402.

Budaev, S. V. 1997. 'Personality' in the guppy (*Poecilia reticulata*): a correlational study of exploratory behaviour and social tendency. *Journal of Comparative Psychology* **111**, 399–411.

Budaev, S. V. and Zhuikov, A. Y. 1998. Avoidance learning and 'personality' in the guppy (*Poecilia reticulata*). *Journal of Comparative Psychology* **112**, 92–94.

Budaev, S. V. and Zworykin, D. D. 2003. Habituation of predator inspection and boldness in the guppy (*Poecilia reticulata*). *Journal of Ichthyology* **43**, S243–S246.

Byholm, P., Ranta, E., Kaitala, V., Linden, H., Saurola, P., and Wikman, M. 2002. Resource availability and goshawk offspring sex ratio variation: a large-scale ecological phenomenon. *Journal of Animal Ecology* **71**, 994–1001.

Cable, J. and Harris, P. D. 2002. Gyrodactylid developmental biology: historical review, current status and future trends. *International Journal for Parasitology* **32**, 255–280.

Cable, J., Scott, E. C. G., Tinsley, R. C., and Harris, P. D. 2002. Behavior favoring transmission in the viviparous monogenean *Gyrodactylus turnbulli*. *Journal of Parasitology* **88**, 183–184.

Caldwell, J. G. 1995. *Environmental situation assessment, Trinidad and Tobago*. Clearwater, Florida: Vista Research Corporation.

Caro, T. M. 1995. Pursuit-deterrence revisited. *Trends in Ecology and Evolution* **10**, 500–503.

Carvalho, G. R., Shaw, P. W., Magurran, A. E., and Seghers, B. H. 1991. Marked genetic divergence revealed by allozymes among populations of the guppy *Poecilia reticulata* (Poeciliidae), in Trinidad. *Biological Journal of the Linnean Society* **42**, 389–405.

Carvalho, G. R., Shaw, P. W., Hauser, L., Seghers, B. H., and Magurran, A. E. 1996. Artificial introductions, evolutionary change and population differentiation in Trinidadian guppies (*Poecilia reticulata*: Poeciliidae). *Biological Journal of the Linnean Society* **57**, 219–234.

Carvalho, G. R., Oosterhout, C. v., Hauser, L., and Magurran, A. E. 2003. Measuring genetic variation in wild populations: from molecular markers to adaptive traits. In *Genes in the environment* (ed. J. Behringer, R. S. Hails, and C. Godfray), pp. 91–111. Oxford: Blackwell Science.

Caswell, H. and Weeks, D. E. 1986. Two-sex models: chaos, extinction and other dynamic consequences of sex. *American Naturalist* **128**, 707–735.

Chace, F. A. and Hobbs, H. H. 1969. *The freshwater and terrestrial decapod crustaceans of the West Indies with special reference to Dominica*. Washington, D.C.: Smithsonian Institution Press.

Chadee, D. D., Ganesh, R., and Persad, R. C. 1991. Feeding behaviour of the Great Kiskadee, *Pitangus sulphuratus*, on fish in Trinidad, West Indies. *Living World* **1991–92**, 42–43.

Chapman, L. J. and Kramer, D. L. 1991*a*. The consequences of flooding for the dispersal and fate of poeciliid fish in an intermittent tropical stream. *Oecologia* **87**, 299–306.

Chapman, L. J. and Kramer, D. L. 1991*b*. Limnological observations of an intermittent tropical dry forest stream. *Hydrobiologia* **226**, 153–166.

Chapman, T. and Partridge, L. 1996. Sexual conflict as fuel for evolution. *Nature* **381**, 189–190.

Chapman, T., Liddle, L., Kalb, J. M., Wolfner, M. F., and Partridge, L. 1995. Costs of mating in *Drosophila melanogaster* females is mediated by male accessory gland products. *Nature* **373**, 241–244.

Chappell, M. and Odell, J. 2004. Predation intensity does not cause microevolutionary change in maximum speed or aerobic capacity in Trinidadian guppies (*Poecilia reticulata* Peters). *Physiological and Biochemical Zoology* **77**, 27–38.

Charlesworth, B. 1978. Model for evolution of Y chromosomes and dosage compensation. *Proceedings of the National Academy of Science (USA)* **75**, 5618–5622.

Charlesworth, B. 1980. *Evolution in age structured populations*. Cambridge: Cambridge University Press.

Charnov, E. L. 1993. *Life history invariants*. Oxford: Oxford University Press.

Chivers, D. P., Brown, G. E., and Smith, R. J. F. 1995. Familiarity and shoal cohesion in fathead minnows (*Pimephales promelas*): implications for antipredator behaviour. *Canadian Journal of Zoology* **73**, 955–960.

Chiyokubo, T. T. S., Nakajima, M., and Fujio, Y. 1998. Genetic features of salinity tolerance in wild and doemestic guppies (*Poecilia reticulata*). *Aquaculture* **167**, 339–348.

Chung, K. S. 2001. Critical thermal maxima and acclimation rate of the tropical guppy *Poecilia reticuata*. *Hydrobiologia* **462**, 253–257.

Clark, E. and Aronson, L. R. 1951. Sexual behavior in the guppy, *Lebistes reticulatus* (Peters). *Zoologica* **36**, 49–66.

Clark, J. A. and May, R. M. 2002. Taxonomic bias in conservation research. *Science* **297**, 191–192.

Clayton, N. S. 1988. Song learning and mate choice in estrilidid finches raised by two species. *Animal Behaviour* **36**, 1589–1600.

Cody, M. L. 1966. A general theory of clutch size. *Evolution* **20**, 174–184.

Coleman, R. M. and Kutty, V. 2001. The predator of guppies in Trinidad is the pike cichlid *Crenicichla frenata*, not *Crenicichla alta*: a caution about working with cichlids. *Journal of Aquaculture and Aquartic Science* **9**, 89–92.

Comfort, A. 1960. The effect of age on growth-resumption in fish (*Lebistes*) checked by food restriction. *Gerontologia* **4**, 177–186.

Comfort, A. 1961. The longevity and mortality of a fish (*Lebistes reticulatus*, Peters) in captivity. *Gerontologia* **5**, 209–222.

Comfort, A. 1963. Effect of delayed and resumed growth on the longevity of a fish (*Lebistes reticulatus*, Peters) in captivity. *Gerontologia* **8**, 150–156.

Connor, R. C. 1996. Partner preferences in by-product mutualisms and the case of predator insepction in fish. *Animal Behaviour* **51**, 451–454.

Constanz, G. D. 1984. Sperm competition in poeciliid fishes. In *Sperm competition and the evolution of mating systems* (ed. R. L. Smith), pp. 465–475. New York: Academic Press.

Constanz, G. D. 1989. Reproductive biology of poeciliid fishes. In *Ecology and evolution of livebearing fishes (Poeciliidae)* (ed. G. K. Meffe and F. F. Snelson), pp. 33–50. Englewood Cliffs, NJ: Prentice Hall.

Coté, I. M. and Hunte, W. 1989. Male and female choice in the redlip blenny: why bigger is better. *Animal Behaviour* **38**, 78–88.

Courtenay, W. R. and Meffe, G. K. 1989. Small fish in strange places: a reveiw of introduced poeciliids. In *Ecology and evolution of livebearing fishes* (ed. G. K. Meffe and F. F. Snelson), pp. 319–331. Englewood Cliffs, NJ: Prentice Hall.

Coyne, J. A. and Orr, H. A. 1989. Patterns of speciation in *Drosophila*. *Evolution* **43**, 362–381.

Coyne, J. A. and Orr, H. A. 1997. 'Patterns of speciation in *Drosophila*' revisited. *Evolution* **51**, 295–303.

Coyne, J. A. and Orr, H. A. 2004. *Speciation*. Sunderland, MA, USA: Sinauer.

Coyne, J. A. and Price, T. D. 2000. Little evidence for sympatric speciation in island birds. *Evolution* **54**, 2166–2171.

Crispo, E., Hendry, A. P., Bentzen, J. J., Reznick, D. N., and Kinnison, M. T. The relative influences of divergent natural selection and physical barriers to dispersal on gene flow in guppies. In review.

Croft, D. P., Albanese, B., Arrowsmith, B. J., Botham, M., Webster, M., and Krause, J. 2003*a*. Sex-biased movement in the guppy (*Poecilia reticulata*). *Oecologia* **137**, 62–68.

Croft, D. P., Arrowsmith, B. J., Bielby, J., Skinner, K., White, E., Couzin, I. D. *et al.* 2003*b*. Mechanisms underlying shoal composition in the Trinidadian guppy, *Poecilia reticulata*. *Oikos* **100**, 429–438.

Croft, D. P., Arrowsmith, B. J., Webster, M., and Krause, J. 2004a. Intra-sexual preferences for familiar fish in male guppies. *Journal of Fish Biology* **64**, 279–283.

Croft, D. P., Botham, M. S., and Krause, J. 2004b. Is sexual segregation in the guppy, *Poecilia reticulata*, consistent with the predation risk hypothesis? *Environmental Biology of Fishes* **71**, 127–133.

Croft, D. P., Krause, J., and James, R. 2004*c*. Social networks in the guppy (*Poecilia reticulata*). Proceedings of the Royal Society of London, Series B.

Darwin, C. 1859. *On the origin of species*. London: John Murray.

Darwin, C. 1871. *The descent of man, and selection in relation to sex*. London: John Murray.

Davis, C. C. 1968. Quantitative feeding and weight change in *Poecilia reticulata*. *Transactions of the American Fisheries Society* **97**, 22–27.

Day, F. 1880. Instincts and emotions in fish. *Linnean Journal—Zoology* **XV**, 31–58.

Day, R. L., MacDonald, T., Brown, C., Laland, K. N., and Reader, S. M. 2001. Interactions between shoal size and conformity in guppy social foraging. *Animal Behaviour* **62**, 917–925.

De Filippi, F. 1861. Note Zoologiche. IV. 'Lebistes' nuovo genere di pesce della famiglia dei Ciprinodonti. *Arch. Zool. Anat. Fisiol. (Genova)* **1861**, 69–70.

De La Vega-Salazar, M. Y., Avila-Luna, E., and Macias-Garcia, C. 2003*a*. Ecological evaluation of local extinction: the case of two genera of endemic Mexican fish, *Zoogoneticus* and *Skiffia*. *Biodiversity and Conservation* **12**, 2043–2056.

De La Vega-Salazar, M. Y., Avila-Luna, E. G., and Macías-Garcia, C. 2003*b*. Threatened fishes of the world: *Zoogoneticus tequila* Webb and Miller 1998 (Goodeidae). *Environmental Biology of Fishes* **68**, 14–14.

De Santi, A., Bisazza, A., Cappelletti, M., and Vallortigara, G. 2000. Prior exposure to a predator influences lateralization of cooperative predator inspection in the guppy, *Poecilia reticulata*. *Italian Journal of Zoology* **67**, 175–178.

Demerec, M. 1928. A possible explanation for Winge's findings in *Lebistes reticulatus*. *American Naturalist* **62**, 90–94.

Diamond, J. M. 1989. Overview of recent extinctions. In *Conservation for the twenty-first century* (ed. D. Western and M. C. Pearl), pp. 37–41. Oxford: Oxford University Press.

Dill, L. M., Hedrick, A. V., and Fraser, A. 1999. Male mating strategies under predation risk: do females call the shots. *Behavioral Ecology* **10**, 452–461.

Doi, T. 1969. Experimental study on the feeding of the guppy *Poecilia reticulata*. *Japanese Journal of Ecology* **19**, 62–66.

Dominey, W. J. 1983. Mobbing in colonially nesting fishes, especially the blue-gill *Lepomis macrochirus*. *Copeia* **1983**, 1086–1088.

Dosen, L. D. and Montgomerie, R. 2004. Female size influences mate preferences of male guppies. *Ethology* **110**, 245–255.

Douglas, M. E. and Endler, J. A. 1982. Quantitative matrix comparisons in ecological and evolutionary investigations. *Journal of Theoretical Biology* **99**, 777–795.

Dowling, T. E. and Moore, W. S. 1986. Absence of population subdivision in the common shiner, *Notropis cornutus* (Cyprinidae). *Environmental Biology of Fishes* **15**, 151–155.
Dugatkin, L. A. 1988. Do guppies play tit for tat during predator inspection visits? *Behavioral Ecology and Sociobiology* **25**, 395–399.
Dugatkin, L. A. 1991*a*. Dynamics of the Tit for Tat strategy during predator inspection in the guppy (*Poecilia reticulata*). *Behavioral Ecology and Sociobiology* **29**, 127–132.
Dugatkin, L. A. 1991*b*. Predator inspection, tit-for-tat and shoaling: a comment on Masters and Waite. *Animal Behaviour* **41**, 898–900.
Dugatkin, L. A. 1992*a*. Sexual selection and imitation: females copy the mate choice of others. *American Naturalist* **139**, 1384–1389.
Dugatkin, L. A. 1992*b*. Tendency to inspect predicts mortality risk in the guppy (*Poecilia reticulata*). *Behavioral Ecology* **3**, 124–127.
Dugatkin, L. A. 1996*a*. The interface between culturally based preferences and genetic preferences: female mate choice in *Poecilia reticulata*. *Proceedings of the National Academy of Science (USA)* **93**, 2770–2773.
Dugatkin, L. A. 1996*b*. Tit for Tat, by-product mutualism and predator inspection: a reply to Connor. *Animal Behaviour* **51**, 455–457.
Dugatkin, L. A. 1997. *Cooperation among animals*. Oxford: Oxford University Press.
Dugatkin, L. A. 1998*a*. A comment on Lafleur et al.'s re-evaluation of mate choice copying in guppies. *Animal Behaviour* **56**, 513–514.
Dugatkin, L. A. 1998*b*. Genes, copying, and female choice: shifting thresholds. *Behavioral Ecology* **9**, 323–327.
Dugatkin, L. A. and Alfieri, M. 1991*a*. Guppies and the Tit for Tat strategy: preference based on past interaction. *Behavioral Ecology and Sociobiology* **28**, 243–246.
Dugatkin, L. A. and Alfieri, M. 1991*b*. Tit for Tat in guppies: the relative nature of cooperation and defection during predator inspection. *Evolutionary Ecology* **5**, 300–309.
Dugatkin, L. A. and Alfieri, M. 1992. Interpopulational differences in the use of the Tit-for-Tat strategy during predator inspection in the guppy, *Poecilia reticulata*. *Evolutionary Ecology* **6**, 519–526.
Dugatkin, L. A. and Alfieri, M. S. 2003. Boldness, behavioral inhibition and learning. *Ethology, Ecology and Evolution* **15**, 43–49.
Dugatkin, L. A. and Godin, J.-G. J. 1992*a*. Predator inspection, shoaling and foraging under predation hazard in the Trinidadian guppy, *Poecilia reticulata*. *Environmental Biology of Fishes* **34**, 265–276.
Dugatkin, L. A. and Godin, J.-G. J. 1992*b*. Prey approaching predators—a cost–benefit perspective. *Annales Zoologici Fennici* **29**, 233–252.
Dugatkin, L. A. and Godin, J.-G. J. 1992*c*. Reversal of female mate choice by copying in the guppy (*Poecilia reticulata*). *Proceedings of the Royal Society of London, Series B* **249**, 179–184.
Dugatkin, L. A. and Godin, J.-G. J. 1993. Female mate copying in the guppy (*Poecilia reticulata*)—age-dependent effects. *Behavioral Ecology* **4**, 289–292.
Dugatkin, L. A. and Godin, J.-G. J. 1998*a*. Effects of hunger on mate choice copying in the guppy. *Ethology* **104**, 194–202.
Dugatkin, L. A. and Godin, J.-G. J. 1998*b*. How females choose their mates. *Scientific American* **278**, 56–61.
Dugatkin, L. A. and Mesterton-Gibbons, M. 1996. Cooperation among unrelated individuals: reciprocal altruism, by-product mutualism and group selection in fishes. *Biosystems* **37**, 19–30.

Dugatkin, L. A. and Sargent, R. C. 1994. Male–male association patterns and female proximity in the guppy, *Poecilia reticulata. Behavioral Ecology and Sociobiology* **35**, 141–145.

Dugatkin, L. A. and Sih, A. 1995. Behavioral ecology and the study of partner choice. *Ethology* **99**, 265–277.

Dugatkin, L. A. and Wilson, D. S. 2000. Assortative interactions and the evolution of cooperation during predator inspection in guppies (*Poecilia reticulata*). *Evolutionary Ecology Research* **2**, 761–767.

Dugatkin, L. A., Lucas, J. S., and Godin, J.-G. J. 2002. Serial effects of mate-choice copying in the guppy (*Poecilia reticulata*). *Ethology, Ecology and Evolution* **14**, 45–52.

Dugatkin, L. A., Druen, M. W., and Godin, J.-G. J. 2003. The disruption hypothesis does not explain mate-choice copying in the guppy *(Poecilia reticulata*). *Ethology* **109**, 67–76.

Dussault, G. V. and Kramer, D. L. 1981. Food and feeding behaviour of the guppy, *Poecilia reticulata* (Pisces: Poeciliidae). *Canadian Journal of Zoology* **59**, 684–701.

Dzikowski, R., Hulata, G., Karplus, I., and Harpaz, S. 2001. Effect of temperature and dietary l-carnitine supplementation on reproductive performance of female guppy (*Poecilia reticulata*). *Aquaculture* **199**, 323–332.

Dzikowski, R., Hulata, G., Harpaz, S., and Karplus, I. 2004. Inducible reproductive plasticity of the guppy *Poecilia reticulata* in response to predation cues. *Journal of Experimental Zoology A—Comparative Experimental Biology* **301A**, 776–782.

Eakley, A. L. and Houde, A. E. 2004. Possible role of female discrimination against 'redundant' males in the evolution of colour pattern polymorphism in guppies. *Proceedings of the Royal Society of London, Series B* **271**, S299–S301.

Elliot, J. M. 1994. *Quantitative ecology and the brown trout*. Oxford: Oxford University Press.

Elton, C. S. 1958. *The ecology of invasions by animals and plants*. London: Methuen and Co.

Endler, J. A. 1978. A predator's view of animal color patterns. *Evolutionary Biology* **11**, 319–364.

Endler, J. A. 1980. Natural selection on color patterns in *Poecilia reticulata. Evolution* **34**, 76–91.

Endler, J. A. 1982. Convergent and divergent effects of natural selection on color patterns in two fish faunas. *Evolution* **36**, 178–188.

Endler, J. A. 1983. Natural and sexual selection on color patterns in poeciliid fishes. *Environmental Biology of Fishes* **9**, 173–190.

Endler, J. A. 1987. Predation, light intensity and courtship behaviour in *Poecilia reticulata* (Pisces: Poeciliidae). *Animal Behaviour* **35**, 1376–1385.

Endler, J. A. 1988. Sexual selection and predation risk in guppies. *Nature* **332**, 593–594.

Endler, J. A. 1991. Variation in the appearance of guppy colour patterns to guppies and their predators under different visual conditions. *Vision Research* **31**, 587–608.

Endler, J. A. 1992*a*. Genetic heterogeneity and ecology. In *Genes in ecology* (ed. R. J. Berry, T. J. Crawford, and G. M. Hewitt), pp. 315–334. Oxford: Blackwell.

Endler, J. A. 1992*b*. Signals, signal conditions and the direction of evolution. *American Naturalist* **139**, S125–S153.

Endler, J. A. 1993. Some general comments on the evolution and design of animal communication systems. *Philosophical Transactions of the Royal Society of London, Series B* **340**, 215–225.

Endler, J. A. 1995. Multiple-trait coevolution and environmental gradients in guppies. *Trends in Ecology and Evolution* **10**, 22–29.

Endler, J. A. 2000. Adaptive genetic variation in the wild. In *Adaptive genetic variation in the wild* (ed. T. A. Mousseau, B. Sinervo and J. Endler), pp. 251–260. New York: Oxford University Press.

Endler, J. A. and Basolo, A. L. 1998. Sensory ecology, receiver biases and sexual selection. *Trends in Ecology and Evolution* **13**, 415–420.

Endler, J. A. and Houde, A. E. 1995. Geographic variation in female preferences for male traits in *Poecilia reticulata*. *Evolution* **49**, 456–468.

Endler, J. A., Basolo, A., Glowacki, S., and Zerr, J. 2001. Variation in response to artificial selection for light sensitivity in guppies (*Poecilia reticulata*). *American Naturalist* **158**, 36–48.

Evans, J. P. 2000. Male mating strategies and sperm competition in the Trinidadian guppy, pp. 106: PhD thesis: University of St Andrews.

Evans, J. P. and Magurran, A. E. 1999*a*. Geographic variation in sperm production by Trinidadian guppies. *Proceedings of the Royal Soiciety of London, Series B* **266**, 2083–2087.

Evans, J. P. and Magurran, A. E. 1999*b*. Male mating behaviour and sperm production characteristics under varying sperm competition risk in guppies. *Animal Behaviour* **58**, 1001–1006.

Evans, J. P. and Magurran, A. E. 2000. Multiple benefits of multiple mating in guppies. *Proceedings of the National Academy of Sciences (USA)* **97**, 10074–10076.

Evans, J. P. and Magurran, A. E. 2001. Patterns of sperm precedence and predictors of paternity in the Trinidadian guppy. *Proceedings of the Royal Society of London, Series B* **268**, 719–724.

Evans, J. P., Kelley, J. L., Ramnarine, I. W., and Pilastro, A. 2002*a*. Female behaviour mediates male courtship under predation risk in the guppy (*Poecilia reticulata*). *Behavioral Ecology and Sociobiology* **52**, 496–502.

Evans, J. P., Pitcher, T. E., and Magurran, A. E. 2002*b*. The ontogeny of courtship, colour and sperm production in guppies. *Journal of Fish Biology* **60**, 495–498.

Evans, J. P., Pilastro, A., and Ramnarine, I. W. 2003*a*. Sperm transfer through forced matings and its evolutionary implications in natural guppy (*Poecilia reticulata*) populations. *Biological Journal of the Linnean Society* **78**, 605–612.

Evans, J. P., Zane, L., Francescato, S., and Pilastro, A. 2003*b*. Directional postcopulatory sexual selection revealed by artificial insemination. *Nature* **421**, 360–363.

Evans, J. P., Bisazza, A., and Pilastro, A. 2004*a*. Female mating preferences for colourful males in a population of guppies subject to high predation. *Journal of Fish Biology* **65**, 1154–1159.

Evans, J. P., Kelley, J. L., Bisazza, A., Finazzo, E., and Pilastro, A. 2004*b*. Sire attractiveness influences offspring performance in guppies. *Proceedings of the Royal Society of London, Series B* **271**, 1035–2042.

Fajan, A. and Breden, F. 1992. Mitochondrial DNA sequence variation among natural populations of the Trinidad guppy, *Poecilia reticulata*. *Evolution* **46**, 1457–1465.

FAO. 1997. *Database on introduced aquatic species*. Rome: FAO.

Farr, J. A. 1975. The role of predation in the evolution of natural populations of the guppy, *Poecilia reticulata* (Pisces: Poeciliidae). *Evolution* **29**, 151–158.

Farr, J. A. 1976. Social facilitation of male sexual behavior, intrasexual competition and sexual selection in the guppy, *Poecilia reticulata* (Pisces: Poeciliidae). *Evolution* **30**, 707–717.

Farr, J. A. 1977. Male rarity or novelty, female choice behavior, and sexual selection in the guppy, *Poecilia reticulata* Peters (Pisces: Poeciliidae). *Evolution* **31**, 162–168.

Farr, J. A. 1980*a*. The effects of sexual experience and female receptivity on courtship-rape decisions in male guppies, *Poecilia reticulata* (Pisces: Poeciliidae). *Animal Behaviour* **28**, 1195–1201.

Farr, J. A. 1980*b*. Social behavior patterns as determinants of reproductive success in the guppy, *Poecilia reticulata* Peters (Pisces: Poeciliidae). An experimental study of the effect pf intermale competition, female choice and sexual selection. *Behaviour* **74**, 38–91.

Farr, J. A. 1981. Biased sex ratios in laboratory strains of guppies, *Poecilia reticulata*. *Heredity* **47**, 237–248.

Farr, J. A. 1983. The inheritance of quantitative fitness traits in guppies *Poecilia reticulata* (Pisces: Poeciliidae). *Evolution* **37**, 1193–1209.

Ferguson, M. M. and Noakes, D. L. G. 1980. Social grouping and genetic variation in common shiners, *Notropus cornutus* (Pisces, Cyprinidae). *Environmental Biology of Fishes* **6**, 357–360.

ffrench, R. 1992. *Birds of Trinidad and Tobago*. London: Christopher Helm.

Field, K. L. and Waite, T. A. 2004. Absence of female conspecifics induces homosexual behaviour in male guppies. *Animal Behaviour* **68**, 1381–1389.

Fisher, R. A. 1914. The evolution of sexual preference. *Eugenics Review* **7**, 184–192.

Fisher, R. A. 1930. *The genetical theory of natural selection*. Oxford: Oxford University Press.

Fisher, R. A. 1958. *The genetical theory of natural selection*. New York: Dover.

FitzGerald, G. J. and Morrisette, J. 1992. Kin recognition and choice of shoal mates by threespine sticklebacks. *Ethology, Ecology and Evolution* **4**, 273–283.

Folstad, I. and Karter, A. J. 1992. Parasites, bright males and the immunocompetence handicap. *American Naturalist* **139**, 603–622.

Foo, C. L., Dinesh, K. R., Lim, T. M., Chan, W. K., and Phang, V. P. E. 1995. Inheritance of RAPD markers in the guppy fish, *Poecilia reticulata*. *Zoological Science* **12**, 535–541.

Fraser, D. F. and Gilliam, J. F. 1987. Feeding under predation hazard: response of the guppy and Hart's rivulus from sites with contrasting predation hazard. *Behavioral Ecology and Sociobiology* **21**, 203–209.

Fraser, D. F. and Gilliam, J. F. 1992. Nonlethal impacts of predator invasion: facultative suppression of growth and reproduction. *Ecology* **73**, 959–970.

Fraser, D. F., Giliam, F. J., and Yip Hoi, T. 1995. Predation as an agent of population fragmentation in a tropical watershed. *Ecology* **76**, 1461–1472.

Fraser, D. F., Gilliam, J. F., MacGowan, M. P., Arcaro, C. M., and Guillozet, P. H. 1999. Habitat quality in a hostile river corridor. *Ecology* **80**, 597–607.

Fraser, D. F., Gilliam, J. F., Akkara, J. T., Albanese, B. W., and Snider, S. B. 2004. Night feeding by guppies under predator release: effects on growth and daytime courtship. *Ecology* **85**, 312–319.

Fuiman, L. A. and Magurran, A. E. 1994. Development of predator defenses in fishes. *Reviews in Fish Biology and Fisheries* **4**, 145–183.

Fujio, Y. and Macaranas, J. M. 1989. Detection of a null allele for MDH isozymes in the guppy (*Poecilia reticulata*), with special reference to sex-linked inheritance. *Japanese Journal of Genetics* **64**, 347–354.

Fujio, Y. and Nakajima, M. 1992. Estimation of genetic load in guppy population. *Nippon Suisan Gakkaishi* **58**, 1603–1605.

Fujio, Y., Nakajima, M., and Hagahama, Y. 1990. Detection of a low temperature-resistant gene in the guppy (*Poecilia reticulata*), with reference to sex-linked inheritance. *Japanese Journal of Genetics* **65**, 201–207.

Fujio, Y., Nakajima, M., and Nomura, G. 1995. Selection response on thermal resistance of the guppy *Poecilia reticulata*. *Fisheries Science* **61**, 731–734.

Futuyma, D. J. 1998. *Evolutionary biology*. Sunderland, MA: Sinauer.

Gabor, C. R. and Halliday, T. R. 1997. Sequential mate choice by multiply mating smooth newts: females become more choosy. *Behavioral Ecology* **8**, 162–166.

Gadgil, M. and Bossert, W. 1970. Life historical consequences of natural selection. *American Naturalist* **104**, 1–24.

Gamble, S., Lindholm, A. K., Endler, J. A., and Brooks, R. 2003. Environmental variation and the maintenance of polymorphism: the effect of ambient light spectrum on mating behaviour and sexual selection in guppies. *Ecology Letters* **6**, 463–472.

Gavrilets, S., Arnqvist, G., and Friberg, U. 2001. The evolution of female mate choice by sexual conflict. *Proceedings of the Royal Society of London, Series B* **268**, 531–539.

Geodakyan, V. A. and Kosobutskii, V. I. 1969. Nature of the feedback mechanism of sex regulation. *Genetika* **5**, 119–126.

Geodakyan, V. A., Kosobutskii, V. I., and Bileva, D. S. 1967. Regulation of sex ratio by negative feedback. *Genetika* **3**, 152–163.

Gerking, S. D. 1994. *Feeding ecology of fish*. San Diego: Academic Press.

Gerlach, G., Schardt, U., Eckmann, R., and Meyer, A. 2001. Kin-structured subpopulations in Eurasian perch (*Perca fluviatilis* L.). *Heredity* **86**, 213–221.

Gerlai, R. 2003. Zebra fish: an uncharted behavior genetic model. *Behavior Genetics* **33**, 461–468.

Ghalambor, C. K., Reznick, D. N., and Walker, J. A. 2004. Constraints on adaptive evolution: the functional trade-off between reproduction and fast-start swimming performance in the Trinidadian guppy *(Poecilia reticulata)*. *American Naturalist* **164**, 38–50.

Gibson, M. B. 1954. Upper lethal temperature relations of the guppy *Lebistes reticulatus*. *Canadian Journal of Zoology* **32**, 393–407.

Gibson, M. B. and Hirst, B. 1955. The effect of salinity and temperature on the pre-adult growth of guppies. *Copeia* **1955**, 241–243.

Gilliam, J. F., Fraser D. F., and Alkins Koo, M. 1993. Structure of a tropical stream fish community—a role for biotic interactions. *Ecology* **74**, 1856–1870.

Gilliam, J. F. and Fraser, D. F. 2001. Movement in corridors: enhancement by predation threat, disturbance and habitat structure. *Ecology* **82**, 258–273.

Godin, J.-G. J. 1986. Antipredator function of shoaling in teleost fishes: a selective review. *Naturaliste Canadiene* **113**, 241–250.

Godin, J.-G. J. 1995. Predation risk and alternative mating tactics in male Trinidadian guppies. *Oecologia* **103**, 224–229.

Godin, J.-G. J. and Davis, S. A. 1995*a*. Boldness and predator deterrence—reply. *Proceedings of the Royal Society of London, Series B* **262**, 107–112.

Godin, J.-G. J. and Davis, S. A. 1995*b*. Who dares, benefits predator approach behaviour in the guppy (*Poecilia reticulata*) deters predator pursuit. *Proceedings of the Royal Society of London, Series B* **259**, 193–200.

Godin, J.-G. J. and Briggs, S. E. 1996. Female mate choice under predation risk in the guppy. *Animal Behaviour* **51**, 117–130.

Godin, J.-G. J. and Dugatkin, L. A. 1995. Variability and repeatability of female mating preference in the guppy. *Animal Behaviour* **49**, 1427–1433.

Godin, J.-G. J. and Dugatkin, L. A. 1996. Female mating preferences for bold males in the guppy, *Poecilia reticulata*. *Proceedings of the National Academy of Sciences, USA* **93**, 10262–10267.

Godin, J.-G. J. and McDonough, H. E. 2003. Predator preference for brightly colored males in the guppy: a viability cost for a sexually selected trait. *Behavioral Ecology* **14**, 194–200.

Godin, J.-G. J. and Smith, S. A. 1988. A fitness cost of foraging in the guppy. *Nature* **333**, 69–71.

Godin, J.-G. J., Alfieri, M. S., Hoare, D. J., and Sadowski, J. A. 2003. Conspecific familiarity and shoaling preference in a wild guppy population. *Canadian Journal of Zoology* **81**, 1899–1904.

Gong, A. 1997. The effects of predator exposure on the female choice of guppies from a high predation population. *Behaviour* **134**, 373–389.

Gong, A. and Gibson, R. M. 1996. Reversal of female preference after visual exposure to a predator in the guppy, *Poecilia reticulata. Animal Behaviour* **52**, 1007–1015.

Goodey, W. and Liley, N. R. 1986. The influence of early experience on escape behaviour in the guppy (*Poecilia reticulata*). *Canadian Journal of Zoology* **64**, 885–888.

Goodrich, H. B., Dee, J. E., Flynn, C. M., and Mercer, R. W. 1934. Germ cells and sex differeintiation in *Lebistes reticulatus. Biological Bulletin* **67**, 83–96.

Goodrich, H. B., Jospehson, N. B., Trinkaus, J. P., and Slate, J. M. 1944. The cellular expression of two new genes in *Lebistes reticulatus. Genetics* **29**, 584–592.

Goodwin, T. W. 1984. *The biochemistry of carotenoids*. London: Chapman and Hall.

Gorlick, D. L. 1976. Dominance hierarchies and factors influencing dominance in the guppy *Poecilia reticulata* (Peters). *Animal Behaviour* **24**, 336–346.

Green, J., Corbet, S. A., Watts, E., and Lan, O. B. 1976. Ecological studies on Indonesian lakes. Overturn and restratification of Ranu Lamongan. *Journal of Zoology* **180**, 315–354.

Green, J., Corbet, S. A., Watts, E., and Lan, O. B. 1978. Ecological studies on Indonesian lakes. The montane lakes of Bali. *Journal of Zoology* **186**, 15–38.

Greenberg, L. A., Hernnäs, B., Brönmark, C., Dahl, J., Eklöv, A., and Olsén, K. H. 2002. Effects of kinship on growth and movements of brown trout in field enclosures. *Ecology of Freshwater Fish* **11**, 251–259.

Gregory, P. G. and Howard, D. J. 1994. A postinsemination barrier to fertilization isolates two closely related ground crickets. *Evolution* **48**, 705–710.

Grether, G. F. 2000. Carotenoid limitation and mate preference evolution: a test of the indicator hypothesis in guppies (*Poecilia reticulata*). *Evolution* **54**, 1712–1724.

Grether, G. F., Hudon, J., and Millie, D. 1999. Carotenoid limitation of sexual coloration along an environmental gradient in guppies. *Proceedings of the Royal Society of London, Series B* **266**, 1317–1322.

Grether, G. F., Hudon, J., and Endler, J. A. 2001*a*. Caretonoid scarcity, synthetic pigments and the evolution of sexual coloration in guppies (*Poecilia reticulata*). *Proceedings of the Royal Society of London, Series B* **268**, 1245–1253.

Grether, G. F., Millie, D. F., Bryant, M. J., Reznick, D. N., and Mayea, W. 2001*b*. Rain forest canopy cover, resource availability, and life history evolution in guppies. *Ecology* **82**, 1546–1559.

Grether, G. F., Kasahara, S., Kolluru, G. R., and Cooper, E. L. 2004. Sex-specific effects of carotenoid intake on the immunological response to allografts in guppies (Poecilia reticulata). *Proceedings of the Royal Society of London, Series B* **271**, 45–50.

Grether, G. F., Cummings, M. E., and Hudon, J. 2005. Countergradient variation in the sexual coloration of guppies (Poecilia reticulata): dosopterin synthesis balances carotenoid availability. *Evolution* **59**, 175–188.

Griffiths, S. W. 1996. Sex differences in the trade-off between feeding and mating in the guppy. *Journal of Fish Biology* **48**, 891–898.

Griffiths, S. W. 2003. Learned recognition of conspecifics by fish. *Fish and Fisheries* **4**, 256–268.

Griffiths, S. W. and Armstrong, J. D. 2001. The benefits of genetic diversity outweigh those of kin association in a territorial animal. *Proceedings of the Royal Society of London, Series B* **268**, 1293–1296.

Griffiths, S. W. and Magurran, A. E. 1997*a*. Familiarity in schooling fish: how long does it take to acquire? *Animal Behaviour* **53**, 945–949.

Griffiths, S. W. and Magurran, A. E. 1997*b*. Schooling preferences for familiar fish vary with group size in wild guppy populations. *Proceedings of the Royal Society of London, Series B* **264**, 547–551.

Griffiths, S. W. and Magurran, A. E. 1998. Sex and schooling behaviour in the Trinidadian guppy. *Animal Behaviour* **56**, 689–693.

Griffiths, S. W. and Magurran, A. E. 1999. Schooling decisions in guppies (*Poecilia reticulata*) are based on familiarity rather than kin recognition by phenotype matching. *Behavioral Ecology and Sociobiology* **45**, 437–443.

Grill, C. P. and Rush, V. N. 2000. Analysing spectral data: comparison and application of two techniques. *Biological Journal of the Linnean Society* **69**, 121–138.

Grossman, G. D., Moyle, P. B., and Whitaker, J. O. 1982. Stochasticity in structural and functional characteristics of an Indiana stream fish assemblage: a test of community theory. *American Naturalist* **120**, 423–454.

Günther, A. C. L. G. 1866. *Catalogue of the fishes in the British Museum, Vol. 6*: British Museum (Natural History).

Haldane, J. B. S. 1922. Sex ratio and unisexual sterility in animal hybrids. *Journal of Genetics* **12**, 101–109.

Halliday, T. R. 1983. The study of mate choice. In *Mate choice* (ed. P. Bateson), pp. 3–32. Cambridge: Cambridge University Press.

Hamilton, W. D. 1966. The moulding of senescence by natural selection. *Journal of Theoretical Biology* **12**, 12–45.

Hamilton, W. D. and Zuk, M. 1982. Heritable true fitness and bright birds: a role for parasites? *Science* **218**, 384–387.

Harris, P. D. and Lyles, A. M. 1992. Infections of Gyrodactylus bullatarudis and Gyrodactylus turnbulli on guppies (*Poecilia reticulata*) in Trinidad. *Journal of Parasitology* **78**, 912–914.

Hartley, D. J., Campbell, K. A., and Suthers, R. A. 1989. The acoustic behavior of the fish-catching bat, *Noctilio leporinus*, during prey capture. *Journal of the Acoustical Society of America* **86**, 8–27.

Harvey, P. H. and Bradbury, J. W. 1991. Sexual selection. In *Behavioural ecology: an evolutionary approach (3rd edn.)* (ed. J. R. Krebs and N. B. Davies), pp. 203–233. Oxford: Blackwell.

Haskins, C. P. and Drizba, J. P. 1938. Note on the anomalous inheritance of sex-linked color factors in the guppy. *American Naturalist* **72**, 571–574.

Haskins, C. P. and Haskins, E. F. 1949. The role of sexual selection as an isolating mechanism in three species of poeciliid fishes. *Evolution* **3**, 160–169.

Haskins, C. P. and Haskins, E. F. 1950. Factors governing sexual selection as an isolating mechanism in the poeciliid fish *Lebistes reticulatus*. *Proceedings of the National Academy of Sciences, USA* **36**, 464–476.

Haskins, C. P. and Haskins, E. F. 1951. The inheritance of certain color patterns in wild populations of *Lebistes reticulatus* in Trinidad. *Evolution* **5**, 216–225.

Haskins, C. P. and Haskins, E. F. 1954. Note on a 'permanent' experimental alteration of genetic constitution in a natural population. *Proceedings of the National Academy of Sciences, USA* **40**, 627–635.

Haskins, C. P., Haskins, E. F., McLaughlin, J. J. A., and Hewitt, R. E. 1961. Polymorphism and population structure in *Lebistes reticulatus*, an ecological study. In *Vertebrate Speciation* (ed. W. F. Blair), pp. 320–395. Austin: University of Texas Press.

Haskins, C. P., Young, P., Hewitt, R. E., and Haskins, E. F. 1970. Stabilized heterozygosis of supergenes mediating certain Y-linked colour patterns in populations of *Lebistes reticulatus*. *Heredity* **25**, 575–589.

Hatai, K., Chukanhom, K., Lawavinit, O. A., Hanjavanit, C., Kunitsune, M., and Imai, S. 2001. Some biological characteristics of *Tetrahymean corlossi* isolated from guppy in Thailand. *Fish Pathology* **34**, 195–199.

Hatfield, T. and Schluter, D. 1999. Ecological speciation in sticklebacks: environment-dependent hybrid fitness. *Evolution* **53**, 866–873.

Haubruge, E., Petit, F., and Gage, M. J. G. 2000. Reduced sperm counts in guppies (*Poecilia reticulata*) following exposure to low levels of tributyltin and bisphenol A. *Proceedings of the Royal Society of London, Series B* **267**, 2333–2337.

Haynes, J. L. 1995. Standardized classification of poeciliid development for life-history studies. *Copeia* **1995**, 147–154.

Hedrick, P. W. 2001. Conservation genetics: where are we now? *Trends in Ecology and Evolution* **16**, 629–636.

Hein, R. G. 1996. Mobbing behavior in juvenile French grunts (*Haemulon flavolineatum*). *Copeia* **1996**, 989–991.

Heinrich, W. and Schröder, J. H. 1986. Can males of *Poecilia reticulata* (Pisces: Poeciliidae) discriminate between females of different genetic quality or of different resident time? *Biologisches Zentrablott* **105**, 491–502.

Herbinger, C. M., Doyle, R. W., Taggart, C. T., Lochmann, S. E., Brooker, A. L., Wright, J. M. *et al.* 1997. Family relationships and effective population size in a natural cohort of Atlantic cod (*Gadus morhua*) larvae. *Canadian Journal of Fisheries and Aquatic Sciences* **54** (Suppl. 1), 11–18.

Herdman, E. J., Kelly, C. D., and Godin, J. G. 2004. Male mate choice in the guppy (*Poecilia reticulata*): do males prefer larger females as mates? *Ethology* **110**, 97–111.

Hester, F. J. 1964. Effects of food supply on fecundity in the female guppy *Lebistes reticulatus* (Peters). *Journal of the Fisheries Research Board of Canada* **21**, 757–764.

Hildemann, W. H. and Wagner, E. D. 1954. Intraspecific sperm competition in *Lebistes*. *American Naturalist* **88**, 87–91.

Hoffmann, A. A. 2000. Laboratory and field heritabilities: some lessons from *Drosophila*. In *Adaptive genetic variation in the wild* (ed. T. A. Mousseau, B. Sinervo, and J. Endler), pp. 200–218. New York: Oxford University Press.

Hornaday, K., Alexander, S., and Breden, F. 1994. Absence of repititive DNA sequences associated with sex chromosomes in natural populations of the Trinidad guppy (*Poecilia reticulata*). *Journal of Molecular Evolution* **39**, 431–433.

Hornaday, K., Alexander, S., and Breden, F. 1995. Distribution of a repeated DNA sequence in natural populations of Trinidad guppy (*Poecilia reticulata*). *Copeia* **1995**, 809–817.

Houde, A. E. 1987. Mate choice based upon naturally occurring color-pattern variation in a guppy population. *Evolution* **41**, 1–10.

Houde, A. E. 1988*a*. The effects of female choice and male–male competition on the mating success of male guppies. *Animal Behaviour* **36**, 888–896.

Houde, A. E. 1988*b*. Genetic difference in female choice between two guppy populations. *Animal Behaviour* **36**, 510–516.

Houde, A. E. 1988*c*. Sexual selection in guppies called into question. *Nature* **333**, 711.

Houde, A. E. 1992. Sex-linked heritability of a sexually selected character in a natural populations of *Poecilia reticulata* (Pisces, Poeciliidae). *Heredity* **69**, 229–235.

Houde, A. E. 1993. Evolution by sexual selection—what can population comparisons tell us? *American Naturalist* **141**, 796–803.

Houde, A. E. 1994. Effect of artificial selection on male colour patterns on mating preference of female guppies. *Proceedings of the Royal Society of London, Series B* **256**, 125–130.

Houde, A. E. 1997. *Sex, color and mate choice in guppies*. Princeton: Princeton University Press.

Houde, A. E. and Endler, J. A. 1990. Correlated evolution of female mating preferences and male color patterns in the guppy, *Poecilia reticuata*. *Science* **248**, 1405–1408.

Houde, A. E. and Hankes, M. A. 1997. Evolutionary mismatch of mating preferences and male colour patterns in guppies. *Animal Behaviour* **53**, 343–351.

Houde, A. E. and Torio, A. J. 1992. Effect of parasitic infection on male color patterns and female choice in guppies. *Behavioral Ecology* **3**, 346–351.

Howard, D. J. 1999. Conspecific sperm preference and pollen precedence and speciation. *Annual Review of Ecology and Systematics* **30**, 109–132.

Hua Yue, G. and Orban, L. 2004. Novel microsatellites from the green swordtail (*Xiphophorus hellerii*) also display polymorphism in guppy (*Poecilia reticulata*). *Molecular Ecology Notes* **4**, 474–476.

Hubbs, C. L. 1934. Racial and individual variation in animals, especially fishes. *American Naturalist* **68**, 115–128.

Hudon, J., Grether, G. E., and Millie, D. F. 2003. Marginal differentiation between sexual and general carotenoid pigmentation of guppies (*Poecilia reticuata*) and a possible visual explanation. *Physiological and Biochemical Zoology* **76**, 776–790.

Hughes, K. A., Du, L., Rodd, F. H., and Reznick, D. N. 1999. Familiarity leads to female mate preference for novel males in the guppy, *Poecilia reticulata*. *Animal Behaviour* **58**, 907–916.

Hughes, K. A., Rodd, F. H., and Reznick, D. N. 2005. Genetic and environmental effects on secondary sex traits in guppies (*Poecilia reticulata*). *Journal of Evolutionary Biology* **18**, 35–45.

Hunte, W. 1978. The distribution of freshwater shrimps (Atyidae and Palaemonidae) in Jamaica. *Zoological Journal of the Linnean Society* **64**, 135–150.

Huntingford, F. A. and Wright, P. J. 1989. How sticklebacks learn to avoid dangerous feeding patches. *Behavioral Processes* **19**, 181–189.

Huntingford, F. A. and Wright, P. J. 1993. The development of adaptive variation in predator avoidance in freshwater fishes. In *Behavioural ecology of fishes* (ed. F. A. Huntingford and P. Torricelli), pp. 45–61. Chur, Switzerland: Harwood.

Immelmann, K. 1972. Sexual and other longterm aspects of imprinting in birds and other species. *Advances in the Study of Behavior* **4**, 147–174.

Irwin, D. E. and Price, T. 1999. Sexual imprinting, learning and speciation. *Heredity* **82**, 347–354.

Janetos, A. C. 1980. Strategies of female mate choice: a theoretical analysis. *Behavioral Ecology and Sociobiology* **7**, 107–112.

Jayawardana, J. M. C. K., Edirisinghe, U., and Silva, L. P. 2000. Morphology, biology and predatory efficiency of *Poecilia reticulata* for mosquito larvae. *Tropical Agricultural Research* **12**, 308–315.

Jennions, M. D. and Petrie, M. 1997. Variation in mate choice and mating preferences: a review of causes and consequences. *Biological Reviews* **72**, 283–327.

Jirotkul, M. 1999*a*. Operational sex ratio influences female preferences and male–male competition in guppies. *Animal Behaviour* **58**, 287–294.

Jirotkul, M. 1999*b*. Population density influences male–male competition in guppies. *Animal Behaviour* **58**, 1169–1175.

Jirotkul, M. 2000*a*. Male trait distribution determined alternative mating tactics in guppies. *Journal of Fish Biology* **56**, 1427–1434.

Jirotkul, M. 2000*b*. Operational sex ratio influences the opportunity for sexual selection in guppies. *Journal of Fish Biology* **56**, 739–741.

Johansson, J., Turesson, H., and Persson, A. 2004. Active selection for large guppies, *Poecilia reticulata*, by the pike cichlid, *Crenicichla saxatilis*. *Oikos* **105**, 595–605.

Johnson, J. E. and Hubbs, C. 1989. Status and conservation of poeciliid fishes. In *Ecology and evolution of livebearing fishes* (ed. G. K. Meffe and F. F. Snelson), pp. 301–317. Englewood Cliffs, NJ: Prentice Hall.

Johnson, T. P., Cullum, A. J., and Bennett, A. F. 1998. Partitioning the effects of temperature and kinematic viscosity on the C-start performance of adult fishes. *Journal of Experimental Biology* **201**, 2045–2051.

Jones, J. C. and Reynolds, J. D. 1997. Effects of pollution on reproductive behaviour of fishes. *Reviews in Fish Biology and Fisheries* **7**, 463–491.

Juliano, R. O., Guerrero, R., and Ronquillo, I. 1989. The introduction of exotic aquatic species in the Phillipines. In *Exotic aquatic organisms in Asia*. (ed. S. S. De Silva), pp. 83–90. Manila: Asian Fisheries Society Special Publication 3.

Kadow, P. 1954. An analysis of sexual behavior and reproductive physiology in the guppy, *Lebistes reticulatus* (Peters). New York: New York University.

Karino, K. and Haijima, Y. 2001. Heritibility of male secondary sexual traits in feral guppies in Japan. *Journal of Ethology* **19**, 33–37.

Karino, K. and Matsunaga, J. 2002. Female mate preference is for male total length, not tail length in feral guppies. *Behaviour* **139**, 1491–1508.

Karplus, I., Gottdiener, M., and Zion, B. 2003. Guidance of single guppies (*Poecilia reticulata*) to allow sorting by computer vision. *Aquacultural Engineering* **27**, 177–190.

Kasumyan, A. O. and Nikolaeva, E. V. 1997. Taste preferences of *Poecilia reticulata* (Cypridontiformes). *Journal of Icthyology* **37**, 696–703.

Kawecki, T. J. and Stearns, S. C. 1993. The evolution of life histories in spatially heterogenous environments: optimal reaction norms revisited. *Evolutionary Ecology* **7**, 155–174.

Keenleyside, M. H. A. 1955. Some aspects of the schooling behaviour of fish. *Behaviour* **8**, 183–248.

Keith, P., LeBail, O.-Y., and Planquette, P. 2000. *Atlas des poissons d'eau douce de Guyane*. Paris: Publications scientifiques du M.N.H.N.

Kelley, J. L. 2002. Behavioural consequences of captive breeding in freshwater fishes, pp. 171: PhD thesis: University of St Andrews.

Kelley, J. L. and Magurran, A. E. 2003*a*. Effects of relaxed predation pressure on visual predator recognition in the guppy. *Behavioral Ecology and Sociobiology* **54**, 225–232.

Kelley, J. L. and Magurran, A. E. 2003*b*. Learned predator recognition and antipredator responses in fish. *Fish and Fisheries* **4**, 216–226.

Kelley, J. L., Graves, J. A., and Magurran, A. E. 1999. Familiarity breeds contempt in guppies. *Nature* **401**, 661.

Kelley, J. L., Evans, J. P., Ramnarine, I. W., and Magurran, A. E. 2003. Back to school: can antipredator behaviour in guppies be enhanced through social learning? *Animal Behaviour* **65**, 655–662.

Kelly, C. D. and Godin, J.-G. J. 2001. Predation risk reduces male–male sexual competition in the Trinidadian guppy (*Poecilia reticulata*). *Behavioral Ecology and Sociobiology* **51**, 95–100.

Kelly, C. D., Godin, J.-J. G., and Wright, J. M. 1999. Geographical variation in multiple paternity within natural populations of the guppy (*Poecilia reticulata*). *Proceedings of the Royal Society of London, Series B* **266**, 2403–2408.

Kelly, C. D., Godin, J.-G. J., and Abdallah, G. 2000. Geographical variation in the male intromittent organ of the Trinidadian guppy (*Poecilia reticulata*). *Canadian Journal of Zoology* **78**, 1674–1680.

Kennedy, C. E. J., Endler, J. A., Poynton, S. L., and McMinn, H. 1987. Parasite load predicts mate choice in guppies. *Behavioral Ecology and Sociobiology* **21**, 291–295.

Kenny, J. S. 1989. Hermatypic scleractinian corals of Trinidad. *Studies of the Fauna of Curaçao and other Caribbean Islands* **123**, 83–100.

Kenny, J. S. 1995. *Views from the bridge: a memoir on the freshwater fishes of Trinidad.* St Joseph, Trinidad and Tobago: J.S. Kenny.

Khanum, S. I., Khan, H. R., and Begum, A. 2002. Larvivorous potential of the guppy Poecilia reticulata on *Culex quinquefasciatus* larvae. *Bangladesh Journal of Zoology* **30**, 41–46.

Khoo, G., Lim, K. F., Gan, D. K. Y., Chen, F., Chan, W. K., Lim, T. M. *et al.* 2002. Genetic diversity within and among feral populations and domesticated strains of the guppy (*Poecilia reticulata*) in Singapore. *Marine Biotechnology* **4**, 367–378.

Khoo, G., Lim, M. H., Suresh, H., Gan, D. K. Y., Lim, K. F., Chen, F. *et al.* 2003. Genetic linkage maps of the guppy (*Poecilia reticulata*): assignment of RAPD markers to multipoint linkage groups. *Marine Biotechnology* **5**, 279–293.

Kim, Y.-K. and Ehrman, L. 1998. Developmental isolation and subsequent adult behavior of *Drosophila paulistorum*. IV. Courtship. *Behavior Genetics* **28**, 57–65.

Kim, Y.-K., Ehrman, L., and Koepfer, H. R. 1996. Developmental isolation and subsequent adult behavior of *Drosophila paulistorum*. II. Prior experience. *Behavior Genetics* **26**, 15–25.

Kinnberg, K., Korsgaard, B., and Bjerregaard, P. 2003. Effects of octylphenol and 17b-estradiol on the gonads of guppies (*Poecilia reticulata*) exposed as adults via the water or as embryos via the mother. *Comparative Biochemistry and Physiology C* **134**, 45–55.

Kinnberg, K. and Toft, G. 2003. Effects of estrogenic and antiandrogenic compounds on the testis structure of the adult guppy (*Poecilia reticulata*). *Ecotoxicology and Environmental Safety* **54**, 16–24.

Kobayashi, H. and Iwamatsu, T. 2002. Fine structure of the storage micropocket of spermatazoa in the ovary of the guppy *Poecilia reticulata. Zoological Science* **19**, 545–555.

Kodric-Brown, A. 1985. Female preference and sexual selection for male coloration in the guppy (*Poecilia reticulata*). *Behavioral Ecology and Sociobiology* **17**, 199–205.

Kodric-Brown, A. 1989. Dietary carotenoids and male mating success in the guppy: an environmental component to female choice. *Behavioral Ecology and Sociobiology* **25**, 393–401.

Kodric-Brown, A. 1992. Male dominance can enhance mating success in guppies. *Animal Behaviour* **44**, 165–167.

Kodric-Brown, A. 1993. Female choice of multiple male criteria in guppies—interacting effects of dominance, coloration and courtship. *Behavioral Ecology and Sociobiology* **32**, 415–420.

Kodric-Brown, A. and Brown, J. H. 1984. Truth in advertising: the kinds of traits favoured by sexual selection. *American Naturalist* **124**, 309–323.

Kodric-Brown, A. and Johnson, S. C. 2002. Ultraviolet reflectance patterns of male guppies enhance their attractiveness to females. *Animal Behaviour* **63**, 391–396.

Kodric-Brown, A. and Nicoletto, P. F. 1996. Consensus among females in their choice of mates in the guppy. *Behavioral Ecology Sociobiology* **39**, 395–400.

Kodric-Brown, A. and Nicoletto, P. F. 1997. Repeatability of female choice in the guppy: response to live and videotaped males. *Animal Behaviour* **54**, 369–376.

Kodric-Brown, A. and Nicoletto, P. 2001*a*. Age and experience affect female choice in the guppy (*Poecilia reticulata*). *American Naturalist* **157**, 316–323.

Kodric-Brown, A. and Nicoletto, P. F. 2001*b*. Female choice in the guppy (*Poecilia reticulata*): the interaction between male color and display. *Behavioral Ecology and Sociobiology* **50**, 346–351.

Kolluru, G. R. and Grether, G. F. 2005. The effects of resource availability on alternative mating tactics in guppies (*Poecilia reticulata*). *Behavioral Ecology* **16**, 294–300.

Koops, M. A. and Abrahams, M. V. 1999. Assessing the ideal free distribution: do guppies use aggression as public information about patch quality? *Ethology* **105**, 737–746.

Kozlowski, J. 1992. Optimal allocation of resources to growth and reproduction: implications for age and size at maturity. *Trends in Ecology and Evolution* **7**, 15–19.

Kozlowski, J. and Uchmansky, Y. 1987. Optimal individual growth and reproduction in perenniel species with indeterminate growth. *Evolutionary Ecology* **1**, 214–230.

Krause, J. and Godin, J.-G. J. 1995. Predator preferences for attacking particular prey group sizes—consequences for predator hunting success and prey predation risk. *Animal Behaviour* **50**, 465–473.

Krause, J. and Godin, J.-G. J. 1996. Influence of prey foraging posture on flight behaviour and predation risk: predators take advantage of unwary prey. *Behavioral Ecology* **7**, 264–271.

Krause, J. and Ruxton, G. D. 2002. *Living in groups*. Oxford: Oxford University Press.

Krumholz, L. A. 1948. Reproduction in the western mosquitofish, *Gambusia affinis affinis* (Baird and Girard) and its use in mosquito control. *Ecological Monographs* **18**, 1–43.

Kruuk, L. E., Clutton-Brock, T. H., Slate, J., Pemberton, J. M., Brotherstone, S., and Guinness, F. E. 2000. Heritability of fitness in a wild mammal population. *Proceedings of the National Academy of Science (USA)* **97**, 698–703.

Kuckuck, C. and Greven, H. 1997. Notes on the mechanically stimulated discharge of spermiozeugmata in the guppy, *Poecilia reticulata*. *Z. Fischk.* **4**, 73–88.

Lachlan, R. F., Crooks, L., and Laland, K. N. 1998. Who follows whom? Shoaling preferences and social learning of foraging information in guppies. *Animal Behaviour* **56**, 181–190.

Lack, D. 1944. Ecological aspects of species formation in passerine birds. *Ibis* **86**, 260–286.

Lacy, R. C. 2000. Structure of the VORTEX simulation model for population viability analysis. *Ecological Bulletins- Swedish Natural Science Research Council* **48**, 191–203.

Lafleur, D. L., Lozano, G. A., and Sclafani, M. 1997. Female mate-choice copying in guppies, *Poecilia reticulata*: a re-evaluation. *Animal Behaviour* **54**, 579–586.

Laland, K. N. and Reader, S. M. 1999. Foraging innovation in the guppy. *Animal Behaviour* **57**, 331–340.

Laland, K. N. and Williams, K. 1997. Shoaling generate social learning of foraging information in guppies. *Animal Behaviour* **53**, 1161–1169.

Lande, R., Engen, S., and Saether, B.-E. 2003. *Stochastic population dynamics in ecology and conservation*. Oxford: Oxford University Press.

Law, R. 1979. Optimal life histories under age-specific predation. *American Naturalist* **114**, 399–417.

Licht, T. 1989. Discriminating between hungry and satiated predators: the response of guppies (*Poecilia reticulata*) from high and low predation sites. *Ethology* **82**, 238–243.

Liley, N. R. 1966. Ethological isolating mechanisms in four sympatric species of Poeciliid fishes. *Behaviour Supplement* **13**, 1–197.

Liley, N. R. 1968. The endocrine control of reproductive behaviour in the female guppy *Poecilia reticulata* (Peters). *Animal Behaviour* **16**, 318–331.

Liley, N. R. and Seghers, B. H. 1975. Factors affecting the morphology and behaviour of guppies in Trinidad. In *Function and evolution in behaviour* (ed. G. Baerends, C. Beer, and A. Manning), pp. 92–118. Oxford: Clarendon Press.

Lindholm, A. and Breden, F. 2002. Sex chromosomes and sexual selection in poeciliid fishes. *American Naturalist* **160** (Suppl.), 214–224.

Lindström, J. and Kokko, H. 1998. Sexual reproduction and population dynamics: the role of polygyny and demographic sex differences. *Proceedings of the Royal Society of London, Series B* **265**, 483–488.

Lindström, K. and Ranta, E. 1993. Social preferences by male guppies, *Poecilia reticulata*, based on shoal size and sex. *Animal Behaviour* **46**, 1029–1031.

Long, K. D. and Houde, A. E. 1989. Orange spots as a visual cue for female mate choice in the guppy (*Poecilia reticulata*). *Ethology* **82**, 316–324.

Long, K. D. and Rosenqvist, G. 1998. Changes in male guppy courting distance in response to a fluctuating light environment. *Behavioral Ecology and Sociobiology* **44**, 77–83.

Lorenz, K. 1937. The campanion in the bird's world. *Auk* **54**, 245–273.

Lorenz, K. 1966. *Evolution and modification of behaviour*. London: Methuen.

Lozano, G. A. 1994. Carotenoids, parasites, and sexual selection. *Oikos* **70**, 309–311.

Luyten, P. H. and Liley, N. R. 1985. Geographic variation in the sexual behaviour of the guppy, *Poecilia reticulata* (Peters). *Behaviour* **95**, 164–179.

Luyten, P. H. and Liley, N. R. 1991. Sexual selection and competitive mating success of male guppies (*Poecilia reticulata*) from four Trinidad populations. *Behavioral Ecology and Sociobiology* **28**, 329–336.

Lyles, A. M. 1990. Genetic variation and susceptibility to parasites: *Poecilia reticulata* infected with *Gyrodactylus turnbulli*: Ph.D. thesis: Princeton University.

Lynch, M., Pfrender, M., Spitze, K., Lehman, N., Hicks, J., Allen, D. *et al.* 1999. The quantitative and molecular genetic architecture of a subdivided species. *Evolution* **53**, 100–110.

Lynch, M. and Walsh, B. 1998. *Genetic analysis of quantitative traits*. Sunderland, USA: Sinauer.

López, S. 1998. Acquired resistance affects male sexual display and female choice in guppies. *Proceedings of the Royal Society of London, Series B* **265**, 717–723.

López, S. 1999. Parasitized female guppies do not prefer showy males. *Animal Behaviour* **57**, 1129–1134.

MacArthur, R. H. 1962. Some generalized themes of natural selection. *Proceedings of the National Academy of Science (USA)* **48**, 1893–1897.

MacArthur, R. H. 1972. *Geographical ecology: patterns in the distribution of species*. New York: Harper and Row.

MacArthur, R. H. and Wilson, E. O. 1967. *The theory of island biogeography*. Princeton: Princeton University Press.

Macías-Garcia, C., Saborío, E., and Berea, C. 1998. Does male-biased predation lead to male scarcity in vivaparous fish? *Journal of Fish Biology* **53** (Suppl. A), 104–117.

Macías-García, C., Jimenez, G., and Contreras, B. 1994. Correlational evidence of a sexually-selected handicap. *Behavioral Ecology and Sociobiology* **35**, 253–259.

Magellen, K., Pettersson, L. B., and Magurran, A. E. Quantifying male attractiveness through phenotypic size manipulation in the Trinidadian guppy, *Poecilia reticulata*. *Behavioral Ecology and Sociobiology*. In press.

Magurran, A. E. 1989. Acquired recognition of predator odour in the European minnow (*Phoxinus phoxinus*). *Ethology* **82**, 216–223.

Magurran, A. E. 1990*a*. The adaptive significance of schooling as an antipredator defence in fish. *Annales Zoologici Fennici* **27**, 51–66.

Magurran, A. E. 1990*b*. The inheritance and development of minnow antipredator behaviour. *Animal Behaviour* **39**, 834–842.

Magurran, A. E. 1996. Battle of the sexes. *Nature* **383**, 307.

Magurran, A. E. 1998. Population differentiation without speciation. *Philosophical Transactions of the Royal Society of London, Series B* **353**, 275–286.

Magurran, A. E. 1999. The causes and consequences of geographic variation in antipredator behavior. In *Geographic variation in behavior* (ed. S. A. Foster and J. A. Endler), pp. 139–163. New York: Oxford University Press.

Magurran, A. E. 2001. Sexual confict and evolution in Trinidadian guppies. *Genetica* **112/113**, 463–474.

Magurran, A. E. 2004. *Measuring biological diversity*. Oxford: Blackwell Science.

Magurran, A. E. and Girling, S. L. 1986. Predator model recognition and response habituation in shoaling minnows. *Animal Behaviour* **34**, 510–518.

Magurran, A. E. and Higham, A. 1988. Information transfer across fish shoals under predator threat. *Ethology* **78**, 153–158.

Magurran, A. E. and Macías-Garcia, C. 2000. Sex differences in behaviour as an indirect consequence of mating system. *Journal of Fish Biology* **57**, 839–857.

Magurran, A. E. and Nowak, M. A. 1991. Another battle of the sexes: the consequences of sexual asymmetry in mating costs and predation risk in the guppy, *Poecilia reticulata. Proceedings of the Royal Society of London, Series B* **246**, 31–38.

Magurran, A. E. and Phillip, D. A. T. 2001*a*. Evolutionary implications of large-scale patterns in the ecology of Trinidadian guppies, *Poecilia reticulata. Biological Journal of the Linnean Society* **73**, 1–9.

Magurran, A. E. and Phillip, D. A. T. 2001*b*. Implications of species loss in freshwater fish assemblages. *Ecography* **24**, 645–650.

Magurran, A. E. and Pitcher, T. J. 1987. Provenance, shoal size and the sociobiology of predator evasion behaviour in minnow shoals. *Proceedings of the Royal Society of London, Series B* **229**, 439–465.

Magurran, A. E. and Ramnarine, I. W. 2004. Learned mate recognition and reproductive isolation in guppies. *Animal Behaviour* **67**, 1077–1082.

Magurran, A. E. and Seghers, B. H. 1990*a*. Population differences in predator recognition and attack cone avoidance in the guppy *Poecilia reticulata. Animal Behaviour* **40**, 443–452.

Magurran, A. E. and Seghers, B. H. 1990*b*. Population differences in the schooling behaviour of newborn guppies, *Poecilia reticulata. Ethology* **84**, 334–342.

Magurran, A. E. and Seghers, B. H. 1990*c*. Risk sensitive courtship in the guppy *Poecilia reticulata. Behaviour* **112**, 194–201.

Magurran, A. E. and Seghers, B. H. 1991. Variation in schooling and aggression amongst guppy, *Poecilia reticulata* populations in Trinidad. *Behaviour* **118**, 214–234.

Magurran, A. E. and Seghers, B. H. 1994*a*. A cost of sexual harassment in the guppy, *Poecilia reticulata. Proceedings of the Royal Society of London, Series B* **258**, 89–92.

Magurran, A. E. and Seghers, B. H. 1994*b*. Predator inspection behaviour covaries with schooling tendency amongst wild guppy, *Poecilia reticulata*, populations in Trinidad. *Behaviour* **128**, 121–134.

Magurran, A. E. and Seghers, B. H. 1994*c*. Sexual conflict as a consequence of ecology: evidence from guppy, *Poecilia reticulata*, populations in Trinidad. *Proceedings of the Royal Society of London, Series B* **255**, 31–36.

Magurran, A. E., Seghers, B. H., Carvalho, G. R., and Shaw, P. W. 1992. Behavioral consequences of an artificial introduction of guppies, *Poecilia reticulata*, in N. Trinidad: evidence for the evolution of antipredator behaviour in the wild. *Proceedings of the Royal Society of London, Series B* **248**, 117–122.

Magurran, A. E., Seghers, B. H., Carvalho, G. R., and Shaw, P. W. 1993. Evolution of adaptive variation in antipredator behaviour. *Marine Behaviour and Physiology* **23**, 29–44.

Magurran, A. E., Seghers, B. H., Shaw, P. W., and Carvalho, G. R. 1994. Schooling preferences for familiar fish in the guppy, *Poecilia reticulata. Journal of Fish Biology* **45**, 401–406.

Magurran, A. E., Seghers, B. H., Shaw, P. W., and Carvalho, G. R. 1995. The behavioral diversity and evolution of guppy, *Poecilia reticulata*, populations in Trinidad. *Advances in the Study of Behavior* **24**, 155–202.

Magurran, A. E., Paxton, C. G. M., Seghers, B. H., Shaw, P. W., and Carvalho, G. R. 1996. Genetic divergence, female choice and male mating success in Trinidadian guppies. *Behaviour* **133**, 503–517.

Maitland, P. S. and Campbell, R. N. 1992. *Freshwater fishes of the British Isles*. London: Harper Collins.

Man, S. H. and Hodgkiss, I. J. 1981. *Hong Kong freshwater fishes*. Hong Kong: Wishing Printing Company.

Mann, K. D., Turnell, E. R., Atema, J., and Gerlach, G. 2003. Kin recognition in juvenile zebrafish (*Danio rerio*) based on olfactory cues. *Biological Bulletin* **205**, 224–225.

Martin, F. D. and Hengstbeck, M. F. 1981. Eye colour and aggression in juvenile guppies, *Poecilia reticulata* Peters (Pisces: Poeciliidae). *Animal Behaviour* **29**, 325–331.

Masters, W. and Waite, T. 1990. Tit-for-tat during predator inspection or shoaling? *Animal Behaviour* **39**, 603–604.

Mathis, A., Chivers, D. P., and Smith, R. J. F. 1996. Cultural transmission of predator recognition in fishes: intraspecific and interspecific learning. *Animal Behaviour* **51**, 185–201.

Matthews, I. M. 1998. Mating behaviour and reproductive biology of the guppy, *Poecilia reticulata*, pp. 208: PhD thesis: University of St Andrews.

Matthews, I. M. and Magurran, A. E. 2000. Evidence for sperm transfer during sneaky mating in wild Trinidadian guppies. *Journal of Fish Biology* **56**, 1381–1386.

Matthews, I. M., Evans, J. P., and Magurran, A. E. 1997. Male display rate reveals ejaculate characteristics in the Trinidadian guppy, *Poecilia reticulata. Proceedings of the Royal Society of London, Series B* **264**, 695–700.

Mattingly, H. T. and Butler, M. J. 1994. Laboratory predation on the Trinidadian guppy: implications for the size-selective predation hypothesis and guppy life history evolution. *Oikos* **69**, 54–64.

May, R. M. 1974. *Stability and complexity in model ecosystems*. Princeton: Princeton University Press.

May, R. M. 2002. The future of biological diversity in a crowded world. *Current Science* **82**, 1325–1331.

Mayr, E. 1942. *Systematics and the origin of species*. NY: Colimbia University Press.

Mayr, E. 1963. *Animal species and evolution*. Cambridge, MA, USA: Harvard University Press.

McCune, A. R. and Lovejoy, N. R. 1998. The relative rate of sympatric and allopatric speciation in fishes. In *Endless forms: species and speciation* (ed. D. J. Howard and S. H. Berlocher), pp. 172–185. New York: Oxford University Press.

McGrew, W.C. 2004. Invention and innovation. *Nature* **427**, 679.

McMinn, H. 1990. Effects of the nematode parasite *Cammalanus cotti* on sexual and non-sexual behaviors in the guppy (*Poecilia reticulata*). *American Zoologist* **30**, 245–249.

Medawar, P. B. 1952. *An unsolved problem of biology*. London: H.K. Lewis and Co.

Meffe, G. K. 1991. Life-history changes in eastern mosquito fish (*Gambusia holbrooki*) induced by thermal elevation. *Canadian Journal of Fisheries and Aquatic Sciences* **48**, 60–66.

Meffe, G. K. 1992. Plasticity of life-history characters in eastern mosquitofish (*Gambusia holbrooki*, Poeciliidae) in response to thermal stress. *Copeia* **1992**, 94–102.

Meffe, G. K., Weeks, S. C., Mulvey, M., and Kandl, K. L. 1995. Genetic differences in thermal tolerance of eastern mosquitofish (*Gambusia holbrooki*; Poeciliidae) from ambient and thermal ponds. *Canadian Journal of Fisheries and Aquatic Sciences* **52**, 2704–2711.

Mendelson, T. C. 2003. Sexual isolation evolves faster than hybrid inviability in a diverse and sexually dimorphic genus of fish (Percidae: *Etheostoma*). *Evolution* **57**, 317–327.

Metcalfe, N. B. and Thomson, B. C. 1995. Fish recognize and prefer to shoal with poor competitors. *Proceedings of the Royal Society of London, Series B* **259**, 207–210.

Meyer, A. 1987. Phenotypic plasticity and heterochrony in *Cichlasoma managuense* (Pisces, Cichlidae) and their implications for speciation in cichlid fishes. *Evolution* **41**, 1357–1369.

Meyer, A. 1989. Cost of morphological specialization: feeding performance of the two morphs in the tropically polymorphic cichlid fish, *Cichlasoma citrinellum*. *Oecologia* **80**, 431–436.

Meyer, J. H. and Liley, N. R. 1982. The control of production of a sexual pheromone in the female guppy *Poecilia reticulata*. *Canadian Journal of Zoology* **60**, 1505–1510.

Michod, R. E. 1979. Evolution of life histories in response to age-specific mortality factors. *American Naturalist* **113**, 531–550.

Mikheev, V. N. and Andreev, O. A. 1993. 2-phase exploration of a novel environment in the guppy, *Poecilia reticulata*. *Journal of Fish Biology* **42**, 375–383.

Milinski, M. 1987. Tit for Tat in sticklebacks and the evolution of cooperation. *Nature* **325**, 433–435.

Milinski, M. 1990. Information overload and food selection. In *Behavioural mechanisms of food selection* (ed. R. N. Hughes), pp. 721–737. Berlin: Springer-Verlag.

Milinski, M. 1996. By-product mutualism, Tit-for-Tat reciprocity and cooperative predator insepction—a reply to Connor. *Animal Behaviour* **51**, 458–461.

Milinski, M. and Boltshauser, P. 1995. Boldness and predator deterrence—a critique of Godin and Davis. *Proceedings of the Royal Society of London, Series B* **262**, 103–105.

Milinski, M. and Heller, R. 1978. Influence of a predator on the optimal foraging behaviour of sticklebacks (*Gasterosteus aculeatus* L.). *Nature* **275**, 642–644.

Milinski, M., Külling, D., and Kettler, R. 1990*a*. Do sticklebacks cooperate repeatedly in reciprocal pairs? *Behavioral Ecology and Sociobiology* **27**, 17–21.

Milinski, M., Külling, D., and Kettler, R. 1990*b*. Tit for tat: sticklebacks 'trusting' a cooperating partner. *Behavioral Ecology* **1**, 7–11.

Milinski, M., Lüthis, J. H., Eggler, R., and Parker, G. A. 1997. Cooperation under predation risk: experiments of costs and benefits. *Proceedings of the Royal Society of London, Series B* **264**, 831–837.

Miller, R. R., Williams, J. D., and Williams, J. E. 1989. Extinctions of North American fish during the last century. *Fisheries* **14**, 22–38.

Mirza, R. S., Chivers, D. P., and Godin, J.-G. J. 2001. Brook charr alevins alter timing of nest emergence in response to chemical cues from fish predators. *Journal of Chemical Ecology* **27**, 1775–1785.

Mohammed, T. I., ChangYen, I., and Bekele, I. 1996. Lead pollution in east Trinidad resulting from lead recycling and smelting activities. *Environmental Geochemistry and Health* **18**, 123–128.

Moore, R. A. and Karasek, F. W. 1984. GC/MS identification of organic pollutants in the Caroni River, Trinidad. *International Journal of Environmental and Analytical Chemistry* **17**, 203–221.

Moreau, R. E. 1944. Clutch-size: a comparative study, with special reference to African birds. *Ibis* **86**, 286–347.

Moyle, P. B. and Leidy, R. A. 1992. Loss of biodiversity in aquatic ecosystems: evidence from fish faunas. In *Conservation biology* (ed. P. L. Fiedler and K. J. Subodh), pp. 127–169. London: Chapman and Hall.

Muntz, W. R. A., Partridge, J. C., Williams, S. R., and Jackson, C. 1996. Spectral sensitivity in the guppy (*Poecilia reticulata*) measured using the dorsal light response. *Marine and Freshwater Behaviour and Physiology* **28**, 163–176.

Murdoch, W. W., Avery, S., and Smyth, M. E. B. 1975. Switching in predatory fish. *Ecology* **56**, 1094–1105.

Nabours, R. K. 1927. Polyandry in the grouse locust, *Paratettix texanus* Hancock, with notes on inheritance of acquired characters and telegony. *American Naturalist* **61**, 531–538.

Naish, K.-A., Carvalho, G. R., and Pitcher, T. J. 1993. The genetic structure and microdistribution of shoals of *Phoxinus phoxinus*, the European minnow. *Journal of Fish Biology* **43**, 75–89.

Nakadate, M., Shikano, T., and Yaniguchi, N. 2003. Inbreeding depression and heterosis in various quantitative traits of the guppy, *Poecilia reticulata*. *Aquaculture* **220**, 219–226.

Nakajima, M. and Fujio, Y. 1993. Genetic determination of the growth of the guppy. *Nippon Suisan Gakkaishi* **59**, 461–464.

Nakajima, M. and Taniguchi, N. 2001. Genetics of the guppy and a model for experiment in aquaculture. *Genetica* **111**, 279–289.

Nakajima, M. and Taniguchi, N. 2002. Genetic control of growth in the guppy (*Poecilia reticulata*). *Aquaculture* **204**, 393–405.

Nanda, I., Feichtinger, W., Schmid, M., Schröder, J., H., Zischler, H., and Epplen, J. T. 1990. Simple repetitive sequences are associated with differentiation of the sex chromosomes in the guppy fish. *Journal of Molecular Evolution* **30**, 456–462.

Nanda, I., Schartl, M., Feichtinger, W., Epplen, J. T., and Schmid, M. 1992. Early stages of sex-chromosome differentiation in fish as analysed by simple repititive DNA sequences. *Chromosoma* **101**, 301–310.

Nanda, I., Schartl, M., Epplen, J. T., Feichtinger, W., and Schmid, M. 1993. Primitive sex chromosomes in poeciliid fishes harbor simple repititive DNA sequences. *Journal of Experimental Zoology* **265**, 301–308.

Nazarova, A. V. and Kreslavskii, A. G. 2000. Comparison of sexual behavior of males of guppy *Poecilia reticulata* of different origin. *Journal of Icthyology* **40**, 191–196.

Neff, B. D. and Pitcher, T. E. 2002. Assessing the statistical power of genetic analyses to detect multiple mating in fishes. *Journal of Fish Biology* **61**, 739–750.

Neff, B. D., Pitcher, T. E., and Repka, J. 2002. A Bayesian model for assessing the frequency of multiple mating in nature. *Journal of Heredity* **93**, 406–414.

Nei, M. 1970. Accumulation of nonfunctional genes on sheltered chromosomes. *American Naturalist* **104**, 311–321.

Neill, S. R. S. and Cullen, J. M. 1974. Experiments on whether schooling by their prey affects the hunting behaviour of cephalopod and fish predators. *Journal of Zoology* **172**, 549–569.

Nicholson, A. J. and Bailey, V. A. 1935. The balance of animal populations. Part I. *Proceedings of the Zoological Society of London* **3**, 551–598.

Nicoletto, P. F. 1991. The relationship between male ornamentation and swimming performance in the guppy, *Poecilia reticulata*. *Behavioral Ecology and Sociobiology* **28**, 365–370.

Nicoletto, P. F. 1993. Female sexual response to condition-dependent ornaments in the guppy, *Poecilia reticulata*. *Animal Behaviour* **46**, 441–450.

Nicoletto, P. F. 1995. Offspring quality and female choice in the guppy, *Poecilia reticulata*. *Animal Behaviour* **49**, 377–387.

Nicoletto, P. F. 1996. The influence of water velocity on the display behavior of male guppies, *Poecilia reticulata*. *Behavioral Ecology* **7**, 272–278.

Nicoletto, P. F. 1999. The use of digitally-modified videos to study the function of ornamentation and courtship in the guppy, *Poecilia reticulata*. *Environmental Biology of Fishes* **56**, 333–341.

Nicoletto, P. F. and Kodric-Brown, A. 1999. The relationship among swimming performance, courtship behavior, and carotenoid pigmentation of guppies in four rivers of Trinidad. *Environmental Biology of Fishes* **55**, 227–235.

Nikolaeva, E. V. and Kasumyan, A. O. 2000. Comparative analysis of the taste preferences and behavioral responses of gustatory stimuli in females and males of the guppy *Poecilia reticulata*. *Journal of Icthyology* **40**, 479–484.

Nishibori, M. and Kawata, M. 1993. The effect of visual density on the fecundity of the guppy, *Poecilia reticulata*. *Environmental Biology of Fishes* **37**, 213–217.

Nordell, S. E. 1998. The response of female guppies, *Poecilia reticulata*, to chemical stimuli from injured conspecifics. *Environmental Biology of Fishes* **51**, 331–338.

Nordell, S. E. and Valone, T. J. 1998. Mate choice copying as public information. *Ecology Letters* **1**, 74–76.

Nursall, J. R. 1973. Some behavioural interactions of spottail shiners (*Notropis hudsonius*), yellow perch (*Perca flavescens*) and northern pike (*Esox lucius*). *Journal of the Fisheries Research Board of Canada* **30**, 1161–1178.

O'Donald, P. 1967. A general model of sexual and natural selection. *Heredity* **22**, 499–518.

O'Donald, P. 1973. Models of sexual and natural selection in polygynous species. *Heredity* **31**, 145–156.

O'Donald, P. 1977. Theoretical aspects of sexual selection. *Theoretical Population Biology* **12**, 298–334.

O'Donald, P. 1980. *Genetic models of sexual selection*. Cambridge: Cambridge University Press.

O'Steen, S., Cullim, A. J., and Bennett, A. F. 2002. Rapid evolution of escape ability in Trinidadian guppies (*Poecilia reticulata*). *Evolution* **56**, 776–784.

Odell, J. P. 2002. Evolution of physiological performance in the Trinidadian guppy (*Poecilia reticulata*: Peters), pp. 114: University of California, Riverside.

Odell, J. P. and Chappell, M. A. 2000. Variation in aerobic capacity within and among 'common-garden' populations of Trinidadian guppies (*Poecilia reticulata*) from different predation regimes. *American Zoologist* **40**, 1157.

Odell, J. P., Chappell, M. A., and Dickson, K. A. 2003. Morphological and enzymatic correlates or aerobic and burst performance in different populations of Trinidadian guppies, *Poecilia reticulata*. *Journal of Experimental Biology* **206**, 3707–3718.

Oetting, S., Prove, E., and Bischop, H. J. 1995. Sexual imprinting as a two-stage process: mechansims of information storage and stabilization. *Animal Behaviour* **50**, 393–403.

Ogutu-Ohwayo, R. 1990. The reduction in fish species diversity in Lake Victoria and Kyoga (East Africa) following human exploitation and introduction of non-native fishes. *Journal of Fish Biology* **37** (Suppl. A), 207–208.

Ogutu-Ohwayo, R. 1993. The effects of predation by Nile Perch, *Lates niloticus* L., on the fish of Lake Nabugabo, with suggestions for conservation of endangered endemic cichlids. *Conservation Biology* **7**, 701.

Ojanguren, A. F. and Magurran, A. E. 2004. Uncoupling the links between male mating tactics and female attractiveness. *Proceedings of the Royal Society of London, Series B* **271**, S427–S429.

Ojanguren, A. F., Evans, J. P., and Magurran, A. E. 2005. Multiple mating influences offspring size in guppies. *Journal of Fish Biology*. In press.

Olendorf, R., Reudi, B., and Hughes, K. A. 2004. Primers for 12 polymorphic microsatellite DNA loci for the guppy (*Poecilia reticulata*). *Molecular Ecology Notes* **4**, 668–671.

Oosterhout, C. v., Harris, P. D., and Cable, J. 2003*a*. Marked variation in parasite resistance between two wild populations of the Trinidadian guppy, *Poecilia reticulata* (Pisces: Poeciliidae). *Biological Journal of the Linnean Society* **79**, 645–651.

Oosterhout, C. v., Trigg, R. E., Carvalho, G. R., Magurran, A. E., Hauser, L., and Shaw, P. W. 2003*b*. Inbreeding depression and genetic load of sexually selected traits: how the guppy lost its spots. *Journal of Evolutionary Biology* **16**, 273–281.

Ortaz, M. 1992. Food-habits of fishes in a neotropical mountain river (in Spanish). *Biotropica* **24**, 550–559.

Palumbi, S. R. and Metz, E. C. 1991. Strong reproductive isolation between closely tropical sea urchins (genus *Echinometra*). *Molecular Biology and Evolution* **8**, 227–239.

Parenti, L. R. 1981. A phylogenetic and biogeographic analysis of cyprinodontiform fishes (Teleostei, Atherinomorpha). *Bulletin of the American Museum of Natural History* **168**, 335–557.

Parenti, L. R. and Rauchenberger, M. 1989. Systematic overview of the Poeciliines. In *Ecology and evolution of livebearing fishes (Poeciliidae)* (ed. G. K. Meffe and F. F. Snelson), pp. 3–12. Englewood Cliffs, NJ: Prentice Hall.

Parker, G. A. 1970. Sperm competition and its evolutionary consequences in insects. *Biological Reviews* **45**, 525–567.

Parker, G. A. and Partridge, L. 1998. Sexual conflict and speciation. *Philosophical Transactions of the Royal Society, London, Series B* **353**, 261–274.

Parker, K. M., Hughes, K., Kim, T. J., and Hedrick, P. W. 1998. Isolation and characterization of microsatellite loci from the Gila topminnow (*Poeciliopsis-o-occidentalis*) and their utility in guppies (*Poecilia reticulata*). *Molecular Ecology* **7**, 361–363.

Pavlov, D. S. and Kasumyan, A. O. 2000. Patterns and mechanisms of schooling behaviour in fish: a review. *Journal of Ichthyology* **40** (Suppl. 2), S163–S231.

Paxton, C. G. M. 1994. Genetic and environmental components of behavioural variation within guppy, *Poecilia reticulata*, populations in Trinidad. DPhil thesis: University of Oxford.

Paxton, C. G. M. 1996. Isolation and development of shoaling in two populations of the guppy. *Journal of Fish Biology* **49**, 514–520.

Paxton, C. G. M., Magurran, A. E., and Zschokke, S. 1994. Caudal eyespots on fish predators influence the inspection behaviour of Trinidadian guppies, *Poecilia reticulata*. *Journal of Fish Biology* **44**, 175–177.

Pelabon, C., Borg, A. A., Bjelvenmark, J., Forsgren, E., Barber, I., and Amundsen, T. 2003. Do male two-spotted gobies prefer large fecund females? *Behavioral Ecology* **14**, 787–792.

Pen, I. and Weissing, F. J. 2002. Optimal sex allocation: steps towards a mechanistic process. In *Sex ratios: concepts and research methods* (ed. I. W. C. Hardy), pp. 26–45. Cambridge: Cambridge University Press.

Peters, W. C. H. 1859. Eine neue vom Herrn Jagor im atlantischen Meere gefangene Art der Gattung leptocephalus, und über einige andere neue Fische des Zoologischen Museums. *Monatsb. Akad. Wiss. Berlin*, 411–413.

Pettersson, L. B. and Brönmark, C. 1999. Energetic consequences of an inducible morphological defence in crucian carp. *Oecologia* **121**, 12–18.

Pettersson, L. B., Ramnarine, I. W., Becher, S. A., Mahabir, R., and Magurran, A. E. 2004. Sex ratio dynamics and fluctuating selection pressures in natural populations of the

Trinidadian gyppy, *Poecilia reticulata*. *Behavioral Ecology and Sociobiology* **55**, 461–468.

Peuhkuri, N. and Seppa, P. 1998. Do three-spined sticklebacks group with kin? *Annales Zoologici Fennici* **35**, 21–27.

Philippi, E. 1908. Fortpflanzungsgeschichte der viviparen Teleosteer *Glaridichthys januarius* und *C. decem-maculatus* in ihrem Einfluss auf Lebenweise, maktoskopische and mikropischeAnatomie. *Zoolog. Jahrbücher* **XXXVII**, 1–94.

Phillip, D. A. T. 1993. Reproduction and feeding of the mountian mullet, *Agonostomus monticola*. *Environmental Biology of Fishes* **37**, 47–55.

Phillip, D. A. T. 1998. Biodiversity of freshwater fishes in Trinidad and Tobago, pp. 99: Ph.D. thesis. University of St Andrews.

Phillip, D. A. T. and Ramnarine, I. W. 2001. *An illustrated guide to the freshwater fishes of Trindad and Tobago*. St Augustine, Trinidad and Tobago: University of the West Indies.

Pianka, E. R. 1970. On r- and K-selection. *American Naturalist* **104**, 592–597.

Pianka, E. R. 1974. *Evolutionary ecology*. New York: Harper and Row.

Pilastro, A. and Bisazza, A. 1999. Insemination efficiency of two alternative male mating tactics in the guppy (*Poecilia reticulata*). *Proceedings of the Royal Society of London, Series B* **266**, 1887–1891.

Pilastro, A., Evans, J. P., Sartorelli, S., and Bisazza, A. 2002. Male phenotype predicts insemination success in guppies. *Proceedings of the Royal Society of London, Series B* **269**, 1325–1330.

Pilastro, A., Simonato, M., Bisazza, A., and Evans, J. P. 2004. Cryptic female preference for colorful males in guppies. *Evolution* **58**, 665–669.

Pinckney, G. A. and Anderson, L. E. 1967. Rearing conditions and sociability in *Lebistes reticulatus*. *Psychonomic Science* **9**, 591–591.

Pitcher, T. E. and Evans, J. P. 2001. Male phenotype and sperm number in the guppy (*Poecilia reticulata*). *Canadian Journal of Zoology* **79**, 1891–1896.

Pitcher, T. E., Neff, B. D., Rodd, F. H, and Rowe, L. 2003. Multiple mating and sequential mate choice in guppies: females trade up. *Proceedings of the Royal Society of London, Series B* **270**, 1623–1629.

Pitcher, T. J. 1983. Heuristic definitions of shoaling behaviour. *Animal Behaviour* **31**, 611–613.

Pitcher, T. J. 1986. Functions of shoaling behaviour in teleosts. In *The behaviour of teleost fishes* (ed. T. J. Pitcher), pp. 294–337. Beckenham, Kent: Croom Helm.

Pitcher, T. J. and Parrish, J. K. 1993. Functions of shoaling behaviour in teleosts. In *Behaviour of teleost fishes* (ed. T. J. Pitcher), pp. 363–439. London: Chapman and Hall.

Pitcher, T. J. and Wyche, C. J. 1983. Predator-avoidance behaviour of sand-eel schools: why do schools seldom split? In *Predators and prey in fishes* (ed. D. L. G. Noakes, B. G. Lindquist, G. S. Helfman, and J. A. Ward), pp. 193–204. The Hague: Junk.

Pocklington, R. and Dill, L. 1995. Predation on females or males: who pays for bright male traits. *Animal Behaviour* **49**, 1122–1124.

Pollock, G. and Dugatkin, L. A. 1992. Reciprocity and the emergence of reputation. *Journal of Theoretical Biology* **159**, 25–37.

Pomiankowski, A. and Sheridan, L. 1994. Linked sexiness and choosiness. *Trends in Ecology and Evolution* **9**, 242–244.

Pomiankowski, A., Iwasa, Y., and Nee, S. 1991. The evolution of costly mate preferences I. Fisher and biased mutation. *Evolution* **45**, 1422–1430.

Pomiankowski, A. N. 1987. The costs of choice in sexual selection. *Journal of Theoretical Biology* **128**, 195–218.

Potts, G. W. 1970. The schooling behaviour of *Lutjianus monostigma* in the shallow reef environment of Aldabra. *Journal of Zoology* **161**, 223–235.

Price, C. S. C. 1997. Conspecific sperm precedence in *Drosophila*. *Nature* **388**, 715–719.

Price, C. S. C., Kim, C. H., Dyer, K. A., and Coyne, J. A. 2000. Mechanisms of conspecific sperm precedence in *Drosophila*. *Evolution* **54**, 2028–2037.

Price, C. S. C., Kim, C. H., Gronlund, C. J., and Coyne, J. A. 2001. Cryptic reproductive isolation in the *Drosophila simulans* clade. *Evolution* **55**, 2028–2037.

Price, J. L. 1955. A survey of freshwater fishes of the island of Trinidad. *Journal of the Agricultural Society of Trinidad and Tobago* **Paper number 863**, 1–28.

Pritchard, V. L., Lawrence, J., Butlin, R. K., and Kraise, J. 2001. Shoal choice in zebrafish, *Danio rerio*: the influence of shoal size and activity. *Animal Behaviour* **62**, 1085–1088.

Purser, G. L. 1937. Succession of broods in *Lebistes*. *Nature* **140**, 155.

Purser, G. L. 1938. Reproduction in *Lebistes reticulatus*. *Quarterly Journal of Microscopical Sciences* **8**, 151–157.

Radakov, D. V. 1973. *Schooling in the ecology of fish*. New York: Wiley.

Ranta, E., Lummaa, V., Kaitala, V., and Merilä, J. 2000. Spatial dynamics and adaptive sex ratios. *Ecology Letters* **3**, 30–34.

Reader, S. M. and Laland, K. N. 2000. Diffusion of foraging innovations in the guppy. *Animal Behaviour* **60**, 175–180.

Reader, S. M. and Laland, K. N. (ed.) 2003. *Animal innovations*. Oxford: Oxford University Press.

Reader, S. M., Kendal, J. R., and Laland, K. N. 2003. Social learning of foraging sites and escape routes in wild Trinidadian guppies. *Animal Behaviour* **66**, 729–739.

Reed, D. H. and Frankham, R. 2001. How closely correlated are molecular and quantitative measures of genetic variation? A meta-analysis. *Evolution* **55**, 1095–1103.

Reede, T. 1995. Life history shifts in response to different levels of fish kairomones in *Daphnia*. *Journal of Plankton Research* **17**, 1661–1667.

Regan, C. T. 1906. On the fresh-water fishes of the island of Trinidad, based on the collection, notes and sketches made by Mr. Lechmere Guppy, Junr. *Proceedings of the Zoological Society of London* 1906, Vol. I, 378–393.

Regan, C. T. 1913. A revision of the Cyprinodont fishes of the subfamily Poeciliinae. *Proceedings of the Zoological Society of London* 1913, Vol II, 977–1018.

Reusch, T. B. H., Haberli, M. A., Aeschlimann, P. B., and Milinski, M. 2001. Female sticklebacks count alleles in a strategy of sexual selection explaining MHC polymorphism. *Nature*, 300–301.

Reynolds, J. D. 1993. Should attractive individuals court more—theory and a test. *American Naturalist* **141**, 914–927.

Reynolds, J. D. and Gross, M. R. 1992. Female mate preference enhances offspring growth and reproduction in a fish, *Poecilia reticulata*. *Proceedings of the Royal Society of London, Series B* **250**, 57–62.

Reynolds, J. D., Gross, M. D., and Coombs, M. J. 1993. Environmental conditions and male morphology determine alternative mating behaviour in Trinidadian guppies. *Animal Behaviour* **45**, 145–152.

Reznick, D. 1982*a*. Genetic determination of offspring size in the guppy (*Poecilia reticulata*). *American Naturalist* **120**, 181–188.

Reznick, D. 1982*b*. The impact of predation on life history evolution in Trinidadian guppies: genetic basis of observed life history patterns. *Evolution* **36**, 1236–1250.

Reznick, D. 1992. Measuring the costs of reproduction. *Trends in Ecology and Evolution* **7**, 42–45.

Reznick, D. 1996. Life history evolution in guppies: a model system for the empirical study of adaptation. *Netherlands Journal of Zoology* **46**, 172–190.

Reznick, D. and Endler, J. A. 1982. The impact of predation on life history evolution in Trinidadian guppies (*Poecilia reticulata*). *Evolution* **36**, 160–177.

Reznick, D. and Yang, A. P. 1993. The influence of fluctuating resources on life-history—patterns of allocation and plasticity in female guppies. *Ecology* **74**, 2011–2019.

Reznick, D., Callahan, H., and Llauredo, R. 1996*a*. Maternal effects on offspring quality in poeciliid fishes. *American Zoologist* **36**, 147–156.

Reznick, D., Buckwalter, G., Groff, J., and Elder, D. 2001*a*. The evolution of senescence in natural populations of guppies (*Poecilia reticulata*): a comparative approach. *Experimental Gerontology* **36**, 791–812.

Reznick, D., Butler, M. J., and Rodd, H. 2001*b*. Life-history evolution in guppies. VII. The comparative ecology of high- and low-predation environments. *American Naturalist* **157**, 126–140.

Reznick, D., Bryant, M. J., and Bashey, F. 2002*a*. r- and K-selection revisited: the role of population regulation in life-history evolution. *Ecology* **83**, 1509–1520.

Reznick, D., Ghalambor, C., and Nunney, L. 2002*b*. The evolution of senescence in fish. *Mechanisms of Aging and Development* **123**, 773–789.

Reznick, D. A., Bryga, H., and Endler, J. A. 1990. Experimentally induced life-history evolution in a natural population. *Nature* **346**, 357–359.

Reznick, D. N. 1983. The structure of guppy life histories: the tradeoff between growth and reproduction. *Ecology* **64**, 862–873.

Reznick, D. N. 1989. Life-history evolution in guppies: 2. Repeatability of field observations and the effects of season on life-histories. *Evolution* **43**, 1285–1297.

Reznick, D. N. 1990. Plasticity in age and size at maturity in male guppies (*Poecilia reticulata*): an experimental evaluation of alternative models of development. *Journal of Evolutionary Biology* **3**, 185–203.

Reznick, D. N. 1997. Life history evolution in guppies (*Poecilia reticulata*): Guppies as a model for studying the evolutionary biology of aging. *Experimental Gerontology* **32**, 245–258.

Reznick, D. N. and Bryga, H. 1987. Life-history evolution in guppies (*Poecilia reticulata*). I. Phenotypic and genetic changes in an introduction experiment. *Evolution* **41**, 137–1385.

Reznick, D. N. and Bryga, H. A. 1996. Life-history evolution in guppies (*Poecilia reticulata*: Poeciliidae). V. Genetic basis of parallelism in life histories. *American Naturalist* **147**, 339–359.

Reznick, D. N. and Ghalambor, C. K. 2001. The population ecology of contemporary adaptations: what empirical studies reveal about the conditions that promote adaptive evolution. *Genetica* **112–113**, 183–198.

Reznick, D. N. and Miles, D. B. 1989. Review of life-history patterns in poeciliid fishes. In *Ecology and evolution of livebearing fishes (Poeciliidae)* (ed. G. K. Meffe and F. F. Snelson), pp. 125–148. Englewood Cliffs, NJ: Prentice Hall.

Reznick, D. N., Miles, D. B., and Winslow, S. 1992. Life-history of *Poecilia picta* (Poeciliidae) from the Island of Trinidad. *Copeia* **1992**, 782–790.

Reznick, D. N., Butler, M. J., Rodd, F. H., and Ross, P. 1996*b*. Life-history evolution in guppies (*Poecilia reticulata*) 6. Differential mortality as a mechanism for natural selection. *Evolution* **50**, 1651–1660.

Reznick, D. N., Rodd, F. H., and Cardenas, M. 1996*c*. Life-history evolution in guppies (*Poecilia reticulata*: Poeciliidae). IV Parallelism in life-history phenotypes. *American Naturalist* **147**, 319–338.

Reznick, D. N., Shaw, F. H., Rodd, F. H., and Shaw, R. G. 1997. Evaluation of the rate of evolution in natural populations of guppies (*Poecilia reticulata*). *Science* **275**, 1934–1937.

Reznick, D. N., Bryant, M. J., Roff, D., Ghalambor, C. K., and Ghalambor, D. E. 2004. Effect of extrinsic mortality on the evolution of senescence in fish. *Nature* **431**, 1095–1099.

Richards, G. R. and Chubb, J. C. 1996. Host response to initial and challenge infections, following treatment of *Gyrodactylus bullatarudis* and *G. turnbulli* (Monogenea) on the guppy (*Poecilia reticulata*). *Parasitology Research* **82**, 242–247.

Richards, G. R. and Chubb, J. C. 1998. Longer-term population dynamics of *Gyrodactylus bullatarudis* and *G. turnbulli* (Monogenea) on adult guppies (*Poecilia reticulata*) in 50-I experimental arenas. *Parasitology Research* **84**, 753–756.

Ricker, W. E. 1954. Stock and recruitment. *Journal of the Fisheries Research Board of Canada* **11**, 559–623.

Ricklefs, R. E. and Miller, G. L. 1999. *Ecology*. New York: W.H. Freeman.

Robinson, B. W. and Wilson, D. S. 1994. Character release and displacement in fishes: a neglected literature. *American Naturalist* **144**, 596–627.

Robinson, B. W. and Wilson, D. S. 1995. Experimentally-induced morphological diversity in Trinidadian guppies. *Copeia* **1995**, 294–305.

Rocchetta, G., Vanelli, M. L., and Pancaldi, C. 2000. Analysis of inheritance of growth trajectories in laboratory populations of guppy-fish. *Growth, Development and Aging* **64**, 83–90.

Rodd, F. H. and Reznick, D. N. 1991. Life history evolution in guppies III. The impact of prawn predation on guppy life histories. *Oikos* **62**, 13–19.

Rodd, F. H. and Reznick, D. N. 1997. Variation in the demography of guppy populations: the importance of predation and life histories. *Ecology* **78**, 405–418.

Rodd, F. H. and Sokolowski, M. B. 1995. Complex origins of variation in the sexual behaviour of male Trinidadian guppies, *Poecilia reticulata*: interactions between social environment, heredity, body size and age. *Animal Behaviour* **49**, 1139–1159.

Rodd, F. H., Reznick, D. N., and Sokolowski, M. B. 1997. Phenotypic plasticity in the life history traits of guppies: responses to social environment. *Ecology* **78**, 419–433.

Rodd, F. H., Hughers, K. A., Grether, G. A., and Baril, C. T. 2002. A possible non-sexual origin of mate preference: are male guppies mimicking fruit? *Proceedings of the Royal Society of London, Series B* **269**, 475–481.

Roff, D. A. and Mousseau, T. A. 1987. Quantitative genetics and fitness: lessons from *Drosophila*. *Heredity* **58**, 103–118.

Rose, S. M. 1959. Population control in guppies. *American Midland Naturalist* **62**, 474–481.

Rosen, D. E. and Bailey, R. M. 1963. The poeciliid fishes (Cyprinodontiformes), their structure, zoogeography and stystematics. *Bulletin of the American Museum of Natural History* **126**, 1–176.

Rosenqvist, G. and Houde, A. 1997. Prior exposure to male phenotypes influences mate choice in the guppy, *Poecilia reticulata*. *Behavioral Ecology* **8**, 194–198.

Rosenthal, H. L. 1951. The birth process of the guppy, *Lebistes reticulatus*. *Copeia* **1951**, 304.

Rosenthal, H. L. 1952. Observations of reproduction of the poeciliid *Lebistes reticulatus* (Peters). *Biological Bulletin* **102**, 30–38.

Roubertoux, P. L. 1992. Courtship behavior in the male guppy (*Poecilia reticulata*): a genetic analysis. *International Journal of Comparative Psychology* **5**, 145–163.

Rowe, L. 1992. Convenience polyandry in a water strider: foraging conflicts and female control of copulation frequency and guarding duration. *Animal Behaviour* **44**, 189–202.

RoyalSociety. 2003. *Measuring biodiversity for conservation*. London: The Royal Society.

Rundle, H. D. 2002. A test of ecologically dependent postmating isolation between sympatric sticklebacks. *Evolution* **56**, 322–329.

Russell, S. T. 2004. Evolution of reproductive isolation in the Trinidadian guppy, *Poecilia reticulata*, pp. 142: Ph.D. thesis. University of St Andrews.

Russell, S. T., Kelley, J. L., Graves, J. A., and Magurran, A. E. 2004. Kin structure and shoal composition dynamics in the guppy, *Poecilia reticulata. Oikos* **106**, 520–526.

Russell, S. T., Ramnarine, I. W., and Magurran, A. E. 2005. Genetic determination of sperm from forced copulations between sympatric populations of *Poecilia reticulata* and *P. picta*. *Biological Journal of the Linnean Society*. In press.

Ryan, M. J. and Rand, A. S. 1993. Species recognition and sexual selection as a unitary problem in animal communication. *Evolution* **47**, 647–657.

Sakwinska, O. 1998. Plasticity of *Daphnia magna* in response to temperature and information about a predator. *Freshwater Biology* **39**, 681–687.

Sato, A., Figueroa, F., Ohuigin, C., Reznick, D. N., and Klein, J. 1996. Identification of major histocompatability complex genes in the guppy, *Poecilia reticulata. Immunogenetics* **43**, 38–49.

Schlupp, I., Waschulewski, M., and Ryan, M. J. 1999. Female preferences for naturally occurring novel male traits. *Behaviour* **136**, 519–527.

Schluter, D. 2000. *The ecology of adaptive radiation*. Oxford: Oxford University Press.

Schmalhausen, I. I. 1949. *Factors of evolution*. Philadelphia, PA, USA: Blakiston.

Schmidt, J. 1919*a*. Racial investigations III. Experiments with *Lebistes reticulatus* (Peters) Regan. *Comptes Rendus des travaux du Laboratoire. Carlsberg* **14** (5), 1–8.

Schmidt, J. 1919*b*. Racial studies in fishes II. Experimental investigations with *Lebistes reticulatus* (Peters) Regan. *Journal of Genetics* **3**, 147–154.

Schmidt, J. 1920. Racial investigations IV. The genetic behavior of a secondary sexual character. Comptes rendus des travaux du laboratorie Carlsberg. Serie Psysiologique. **14**, 227.

Schnitzler, H. U., Kalko, E. K. V., Kaipf, I. and Grinnell, A. D. 1994. Fish and echolocation behavior of the greater bulldog bat, *Noctilio leporinus*, in the field. *Behavioral Ecology and Sociobiology* **35**, 327–345.

Schröder, J. 1983. The guppy (*Poecilia reticulata*) as a model for evolutionary studies in genetics, behaviour, and ecology. *Beratungen des Naturwissencscchaftlich-medizinischen Vereins Innsbruck* **70**, 249–279.

Schröder, J. H. 1993. Social behaviour differences between XY and YY males of the guppy, *Poecilia reticulata* Peters. *Behavior Genetics* **23**, 565–565.

Schröder, J. H. and Holzberg, S. 1972. Population genetics of *Lebistes (Poecilia) reticulatus* Peters (Poeciliidae: Pisces). I. Effects of radiation induced mutations on the segregation ratio in posirradiation F2. *Genetics* **70**, 621–630.

Schröder, J. H. and Peters, K. 1988. Differential courtship activity of competing guppy males (*Poecilia reticulata* Peters; Pisces: Poeciliidae) as an indicator for low concentrations of aquatic pollutants. *Bulletin of Environmental Contamination and Toxicology* **40**, 396–404.

Scott, M. E. 1985. Dynamics of challenge infection of *Gyrodactylus bullatarudis* Turnbull (Monogenea) on guppies, *Poecilia reticulata* (Peters). *Journal of Fish Diseases* **8**, 485–503.

Scott, M. E. and Anderson, R. M. 1984. The population dynamics of *Gyrodactylus bullatarudis* (Monogenea) within laboratory populations of the host fish. *Parasitology* **89**, 159–194.

Seger, J. and Brockman, H. J. 1987. What is bet hedging? In *Oxford surveys in evolutionary biology* (ed. P. Harvey and L. Partridge), pp. 182–211. Oxford: Oxford University Press.

Seghers, B. H. 1973. *An analysis of geographic variation in the antipredator adaptations of the guppy, Poecilia reticulata*. Ph.D. thesis: University of British Columbia.

Seghers, B. H. 1974*a*. Geographic variation in the responses of guppies (*Poecilia reticulata*) to aerial predators. *Oecologia* **14**, 93–98.

Seghers, B. H. 1974*b*. Schooling behavior in the guppy (*Poecilia reticulata*): an evolutionary response to predation. *Evolution* **28**, 486–489.

Seghers, B. H. 1992. The rivers of northern Trinidad: conservation of fish communities for research. In *River conservation and management* (ed. P. J. Boon, G. E. Petts and P. Calow), pp. 81–90. Chichester: Wiley.

Seghers, B. H. and Magurran, A. E. 1995. Population differences in the schooling behaviour of the Trinidad guppy, *Poecilia reticulata*—adaptation or constraint? *Canadian Journal of Zoology* **73**, 1100–1105.

Serebov, L. I. 1973. Effect of a current on the intensity of feeding in a certain fish. *Hydrobiologia Journal* **9**, 68–70.

Shaw, E. 1962. The schooling of fishes. *Scientific American* **206**, 128–138.

Shaw, E. 1978. Schooling fishes. *American Scientist* **66**, 166–175.

Shaw, P. W., Carvalho, G. R., Magurran, A. E., and Seghers, B. H. 1991. Population differentiation in Trinidadian guppies (*Poecilia reticulata*)—patterns and problems. *Journal of Fish Biology* **39** (Suppl. A), 203–209.

Shaw, P. W., Carvalho, G. R., Seghers, B. H., and Magurran, A. E. 1992. Genetic consequences of an artificial introduction of guppies (*Poecilia reticulata*) in N. Trinidad. *Proceedings of the Royal Society of London, Series B* **248**, 111–116.

Shaw, P. W., Carvalho, G. R., Magurran, A. E., and Seghers, B. H. 1994. Factors affecting the distribution of genetic variabilty in the guppy, *Poecilia reticulata*. *Journal of Fish Biology* **45**, 875–888.

Sheldon, B. C. 2000. Differential allocation: tests, mechanisms and implications. *Trends in Ecology and Evolution* **15**, 397–402.

Sheridan, L. and Pomiankowski, A. 1997*a*. Female choice for spot asymmetry in the Trinidadian guppy. *Animal Behaviour* **54**, 1523–1529.

Sheridan, L. and Pomiankowski, A. 1997*b*. Fluctuating asymmetry, spot asymmetry and inbreeding depression in the sexual coloration of male guppy fish. *Heredity* **79**, 515–523.

Shettleworth, S. J. 1998. *Cognition, evolution, and behavior*. New York: Oxford University Press.

Shikano, T. and Fujio, Y. 1998. Effects of the mother's environmental salinity on seawater tolerance of newborn guppy, *Poecilia reticulata*. *Fisheries Science* **64**, 10–13.

Shikano, T. and Taniguchi, N. 2002*a*. Heterosis for neonatal survival in the guppy. *Journal of Fish Biology* **60**, 715–725.

Shikano, T. and Taniguchi, N. 2002*b*. Relationships between genetic variation measured by microsatellite DNA markers and a fitness-related trait in the guppy (*Poecilia reticulata*). *Aquaculture* **209**, 77–90.

Shikano, T. and Taniguchi, N. 2002*c*. Using microsatellite and RADPmarkers to estimatethe amount of heterosis in various strain combinations in the guppy (*Poecilia reticulata*) as a fish model. *Aquaculture* **204**, 271–281.

Shikano, T. and Taniguchi, N. 2003. DNA markers for estimation of inbreeding and heterosis in the guppy *Poecilia reticulata*. *Aquaculture Research* **34**, 905–911.

Shikano, T., Nakadate, M., Nakajima, M., and Fujio, Y. 1997. Heterosis and maternal effects in salinity tolerance of the guppy, *Poecilia reticulata*. *Fisheries Science* **63**, 893–896.

Shikano, T., Chiyokubo, T., Nakadate, M., and Fujio, Y. 2000*a*. The relationship between allozyme heterozygosity and salinity tolerance in wild and domestic populations of the guppy (*Poecilia reticulata*). *Aquaculture* **184**, 233–245.

Shikano, T., Nakadate, M., and Fuijo, Y. 2000*b*. An experimental study on strain combinations in salinity tolerance of the guppy, *Poecilia reticulata. Fisheries Science* **66**, 625–632.

Shikano, T., Chiyokubo, T. and Taniguchi, N. 2001*a*. Effect of inbreeding on salinity tolerance in the guppy (*Poecilia reticulata*). *Aquaculture* **202**, 45–55.

Shikano, T., Chiyokubo, T. and Taniguchi, N. 2001*b*. Temporal changes in allele frequency, genetic variation and inbreeding depression in small populations of the guppy, *Poecilia reticulata. Heredity* **86**, 153–160.

Shohet, A. J. and Watt, P. J. 2004. Female association preferences based on olfactory cues in the guppy, *Poecilia reticulata. Behavioral Ecology and Sociobiology* **55**, 363–369.

Sih, A. 1994. Predation risk and the evolutionary ecology of reproductive behaviour. *Journal of Fish Biology* **45**, 111–130.

Sih, A., Bell, A., and Johnson, J. C. 2004. Behavioral syndromes: an ecological and evolutionary overview. *Trends in Ecology and Evolution* **19**, 372–378.

Silliman, R. P. and Gutsell, J. S. 1958. Experimental exploitation of fish populations. *Fisheries Bulletin* **58**, 215–252.

Sirot, E. 2001. Mate-choice copying by females: the advantages of a prudent strategy. *Journal of Evolutionary Biology* **14**, 418–423.

Skelly, D. K. and Werner, E. E. 1990. Behavioral and life historical responses of larval American toads to an odonate predator. *Ecology* **71**, 2313–2322.

Skúlason, S., Snorrason, S. S., and Jónsson, B. 1999. Sympatric morphs, populations and speciation in freshwater fish with emhpasis on arctic charr. In *Evolution of biological diversity* (ed. A. E. Magurran and R. M. May), pp. 70–92. Oxford: Oxford University Press.

Slagsvold, T., Hansen B. T., Johannessen, L. E., and Lifjeld, J. T. 2002. Mate choice and imprinting in birds studies by cross-fostering in the wild. *Proceedings of the Royal Society of London, Series B* **269**, 1449–1455.

Smith, E. J., Partridge, J. C., Parsons, K. N., White, E. M., Cuthill, I. C., Bennett, A. T. D. *et al.* 2002. Ultraviolet vision and mate choice in the guppy (*Poecilia reticulata*). *Behavioral Ecology* **13**, 11–19.

Smith, R. J. F. 1992. Alarm signals in fishes. *Reviews in Fish Biology and Fisheries* **2**, 22–63.

Smuts, B. B. and Smuts, R. W. 1993. Male aggression and sexual coercion of females in nonhuman primates and other mammals: evidence and theoretical implications. *Advances in the Study of Behavior* **22**, 1–63.

Snelson, F. F. 1989. Social and environmental control of life history traits in poeciliid fishes. In *Ecology and evolution of livebearing fishes* (ed. G. K. Meffe and F. F. Snelson), pp. 149–162. Englewood Cliffs, NJ: Prentice Hall.

Southwood, T. R. E., Henderson, P. A., and Woiwod, I. P. 2003. Stability and change over 67 years—the community of heteroptera as caught in a light-trap at Rothamsted, UK. *European Journal of Entomology* **100**, 557–561.

Spurway, H. 1953. Spontaneous parthenogenesis in fish. *Nature* **171**, 437.

Spurway, H. 1957. Hermaphroditism with self-fertilisation and the monthly extrusion of unfertilised eggs in the viviparous fish *Lebistes reticulatus. Nature* **180**, 1248–1251.

Stearns, S. C. 1977. The evolution of life history traits. A critique of the theory and a review of the data. *Annual Review of Ecology and Systematics* **8**, 145–171.

Stearns, S. C. 1992. *The evolution of life histories*. Oxford: Oxford University Press.

Stepanek, O. 1928. Morfologie a biologis genitalnich organu u *Lebistes reticulatus* Peters. *Pub. fac. Sci. Univ. Charles* **79**, 1–30.

Stibor, H. 1992. Predator induced life history traits in a freshwater cladoceran. *Oecologia* **92**, 162–165.

Stockwell, C. A., Hendry, A. P., and Kinnison, M. T. 2003. Contemporary evolution meets conservation biology. *Trends in Ecology and Evolution* **18**, 94–101.

Stolk, A. 1951. Histo-endocrinological analysis of gestation phenomena in the cyprinodont *Lebistes reticulatus* (Peters). *Proceedings of the Academy of Sciences, Amsterdam* **54C**, 550–578.

Stoner, G. and Breden, F. 1988. Phenotypic differentiation in female preference related to geographic variation in male predation risk in the Trinidad guppy (*Poecilia reticulata*). *Behavioral Ecology and Sociobiology* **22**, 285–291.

Strauss, R. E. 1990. Predation and life-history variation in *Poecilia reticulata* (Cyprinodontiformes: Poeciliidae). *Environmental Biology of Fishes* **27**, 121–130.

Suboski, M. D., Bain, S., Carty, A. E., McQuoid, L. M., Seelen, M. I., And Seifert, M. 1990. Alarm reaction in acquisition and social transmission of simulated predator-recognition by zebra danio fish (*Brachydanio rerio*). *Journal of Comparative Pyschology* **104**, 101–112.

Sugita, Y. 1980. Imitative choice behaviour in guppies. *Japan Psychological Research* **22**, 7–12.

Swaney, W., Kendal, J., Capon, H., Brown, C., and Laland, K. N. 2001. Familiarity facilitates social learning of foraging behaviour in the guppy. *Animal Behaviour* **62**, 591–598.

Takahashi, H. 1975. Process of functional sex reversal of the gonad in the female guppy, *Poecilia reticulata*, treated with androgen before birth. *Development, Growth and Differentiation* **17**, 167–175.

Taylor, J. S. and Breden, F. 2000. Slipped-strand mispairing at noncontiguous repeats in *Poecilia reticulat*a: a model for minisatellite birth. *Genetics* **155**, 1313–1320.

Taylor, J. S. and Breden, F. 2002. The inheritance of heteroplasmy in guppies. *Journal of Fish Biology* **60**, 1346–1350.

Taylor, J. S., Durkin, M. H., and Breden, F. 1999*a*. The death of a microsatellite: a phylogenetic perspective on microsatellite interruptions. *Molecular Biology and Evolution* **16**, 567–572.

Taylor, J. S., Sanny, J. S. P., and Breden, F. 1999*b*. Microsatellite allele size homoplasy in the guppy (*Poecilia reticulata*). *Journal of Molecular Evolution* **48**, 245–247.

Tebb, G. and Thoday, J. M. 1956. Reversal of mating preference by crossing strains of *Drosophila melanogaster*. *Nature* **177**, 707.

Templeton, C. N. and Shriner, W. M. 2004. Multiple selection pressures influence Trinidadian guppy (*Poecilia reticulata*) antipredator behavior. *Behavioral Ecology* **15**, 673–678.

Thibault, R. E. and Schultz, R. J. 1978. Reproductive adaptations among viviparous fishes (Cyprinodontiformes: Poeciliidae). *Evolution* **32**, 320–333.

Thilakarante, I. D. S. I. P., Rajapaksha, G., Hewakopara, A., Rajapakse, R. P. V. J., and Faizal, A. C. M. 2003. Parasitic infections in freshwater ornamental fish in Sri Lanka. *Diseases of Aquatic Organisms* **54**, 157–162.

Thomas, J. A., Telfer, M. G., Roy, D. B., Peston, C. D., Greenwood, J. J. D., Asher, J. *et al.* 2004. Comparative losses of British butterflies, birds, and plants and the global extinction crisis. *Science* **303**, 1879–1881.

Thompson, E. A. and Shaw, R. G. 1992. Estimating polygenic models for multivariate data on large pedigrees. *Genetics* **131**, 971–978.

Toft, G. and Baatrup, E. 2001. Sexual characteristics are altered by 4-tert-octylphenol and 17b-estradiol in the adult male guppy (*Poecilia reticulata*). *Ecotoxicology and Environmental Safety* **48**, 76–84.

Toft, G. and Baatrup, E. 2003. Altered sexual characteristics in guppies (*Poecilia reticulata*) exposed to 17 beta-estradiol and 4-tert-octylphenol during sexual development. *Ecotoxicology and Environmental Safety* **56**, 228–237.

Topal, J., Miklosi, A., and Csanyi, V. 1994. Is the testing of TFT strategy during predator inspections possible by use of mirror image—a review and some controls of recent studies. *Acta Biologica Hungarica* **45**, 87–99.

Travis, J. 1989. Ecological genetics of life history traits in poeciliid fishes. In *Ecology and evolution of livebearing fishes* (ed. G. K. Meffe and F. F. Snelson), pp. 185–200. Englewood Cliffs, New Jersey: Prentice Hall.

Trexler, J. C. 1989. Phenotypic plasticity in poeciliid life histories. In *Ecology and evolution of livebearing fishes* (ed. G. K. Meffe and F. F. Snelson), pp. 201–214. Englewood Cliffs, New Jersey: Prentice Hall.

Trexler, J. C. 1997. Resource availability and plasticity in offspring provisioning: embryo nourishment in sailfin mollies. *Ecology* **78**, 1370–1381.

Trexler, J. C. and DeAngelis, D. L. 2003. Resource allocation in offspring provisioning: an evaluation of the conditions favoring the evolution of matrotrophy. *American Naturalist* **162**, 574–585.

Trivers, R. L. and Willard, D. E. 1973. Natural selection of parental ability to vary the sex ratio of offspring. *Science* **191**, 249–263.

Turelli, M., Barton, N. H., and Coyne, J. A. 2001. Theory and speciation. *Trends in Ecology and Evolution* **16**, 330–343.

Turner, C. L. 1947. Viviparity in teleost fishes. *Science Monthly* **65**, 508–518.

Uematsu, T. 1971. Social facilitation in feeding behaviour of the guppy. II. Experimental analysis of mechanisms. *Japanese Journal of Ecology* **21**, 54–67.

Urban-Jezierska, E. 2002. Effect of acidification on feeding in guppy fish (*Lebistes reticulatus* vel *Poecilia reticulata*). *Polish Journal of Ecology* **50**, 25–44.

Van Havre, N. and Fitzgerald, G. J. 1988. Shoaling and kin recognition in the threespine stickleback (*Gasterosteus aculeatus* L.). *Biology of Behaviour* **13**, 190–201.

Vogel, D. and Bleckmann, H. 2000. Behavioral discrimination of water motions caused by moving objects. *Journal of Comparative Physiology A* **186**, 1107–1117.

Vogel, J. L. and Beauchamp, D. A. 1999. Effects of light, prey size, and turbidity on reaction distances of lake trout (Salvelinus namaycush) to salmonid prey. *Canadian Journal of Fisheries and Aquatic Sciences* **56**, 1293–1297.

Vumpurath, S. and Pandian, T. J. 1993*a*. Masculinization of *Poecilia reticulata* by dietary administration of synthetic or natural androgen to gravid females. *Aquaculture* **116**, 83–89.

Vumpurath, S. and Pandian, T. J. 1993*b*. Production of a YY female guppy, *Poecilia reticulata*, by endocrine sex reversal and progeny testing. *Aquaculture* **118**, 183–189.

Warburton, K. and Lees, N. 1996. Species discrimination in guppies: learned responses to visual cues. *Animal Behaviour* **52**, 371–378.

Warren, E. W. 1973*a*. Modification of the response to high density conditions in the guppy, *Poecilia reticulata* (Peters). *Journal of Fish Biology* **5**, 737–752.

Warren, E. W. 1973*b*. The effect of relative denisty upon aspects of the behaviour of the guppy, *Poecilia reticulata*. *Journal of Fish Biology* **5**, 753–765.

Warren, M. L., Burr, B. M., Walsh, S. J., Bart, H. L., Cashner, R. C., Etnier, D. A. *et al.* 2000. Diversity, distribution, and conservation status of the native freshwater fishes of the southern United States. *Fisheries* **25**, 7–31.

Water Resources Management Unit, T. 2002. *Draft national water resources management policy*. Port of Spain: Government of the Republic of Trinidad and Tobago.

Watt, P. J., Shohet, A. J., and Renshaw, K. 2001. Female choice for good genes and sex-biased broods in guppies. *Journal of Fish Biology* **59**, 843–850.

Webb, P. W. 1982. Avoidance responses of fathead minnow to strikes by four teleost predators. *Journal of Comparative Physiology* **147**, 371–378.

Webb, P. W. and Skadsen, J. M. 1980. Strike tactics of *Esox*. *Canadian Journal of Zoology* **58**, 1462–1469.

Webb, S. A., Graves, J. A., Macias-Garcia, C., Magurran, A. E., Foighil, D. O., and Ritchie, M. G. 2004. Molecular phylogeny of the livebearing Goodeidae (Cyprinodontiformes). *Molecular Phylogenetics and Evolution* **30**, 527–544.

Webster, M. S. and Almany, G. R. 2002. Positive indirect effects in a coral reef fish community. *Ecology Letters* **5**, 549–557.

Weetman, D., Atkinson, D., and Chubb, J. C. 1998. Effect of temperature on anti-predator behaviours in the guppy, *Poecilia reticulata*. *Animal Behaviour* **55**, 1361–1372.

Weetman, D., Atkinson, D., and Chubb, J. C. 1999. Water temperature influences the shoaling decisions of guppies, *Poecilia reticulata*, under predation threat. *Animal Behaviour* **58**, 735–741.

Weider, L. J. and Pijanowska, J. 1993. Plasticity of daphnia life histories in response to chemical cues from predators. *Oikos* **67**, 385–392.

Welcomme, R. L. 1988. *International introductions of inland aquatic species*: FAO Fisheries Technical Paper 294.

Wenstrup, J. J. and Suthers, R. A. 1984. Echolocation of moving tragets by the fish-catching bat *Noctilio leporinus*. *Journal of Comparative Physiology* **155**, 75–89.

Westneat, D. F., Walters, A., McCartky, T. M., Hatch, M. I., and Hein, W. K. 2000. Alternative mechanisms of nonindependent mate choice. *Animal Behaviour* **59**, 467–476.

Wheeler, A. C., Merrett, N. R., and Quigley, D. T. G. 2004. Additional records and notes for Wheeler's (1992) List of the common and scientific names of fishes of the British Isles. *Journal of Fish Biology* **65** (Suppl. B), 1–40.

White, E. M., Partridge, J. C., and Church, S. C. 2003. Ultraviolet dermal reflexion and mate choice in the guppy, *Poecilia reticulata*. *Animal Behaviour* **65**, 693–700.

Widianarko, B., Van Gestel, C. A. M., Verweij, R. A., and Van Strallen, N. M. 2000. Associations between trace metals in sediment, water and guppy, *Poecilia reticulata* (Peters), from urban streams of Semarang, Indonesia. *Ecotoxicology and Environmental Safety* **46**, 101–107.

Wilbur, H. M., Tinkle, D. W., and Tilley, S. G. 1970. Environmental certainty, trophic level, and resource availability in life history evolution. *American Naturalist* **108**, 805–816.

Williams, C. B. 1964*a*. *Patterns in the balance of nature*. London: Academic Press.

Williams, G. C. 1957. Pleitropy, natural selection and the evolution of senescence. *Evolution* **11**, 389–411.

Williams, G. C. 1964*b*. Measurement of consociation among fishes and comments on the evolution of schooling. *Publications of the Museum, Michigan State University, Biological Series* **2**, 349–384.

Williams, G. C. 1966. Natural selection, the costs of reproduction, and a refinement of Lack's principle. *American Naturalist* **100**, 687–690.

Wilson, J., Kuehn, R., and Beach, F. 1963. Modifications in the sexual behavior of male rates produced by changing the stimulus female. *Journal of Comparative and Physiological Psychology* **56**, 636–644.

Winemiller, K. O. 1993. Seasonality of reproduction by livebearing fishes in tropical rain forest streams. *Oecologia* **95**, 266–276.

Winemiller, K. O., Leslie, M., and Roche, R. 1990. Phenotypic variation in male guppies from natural inland populations: an additional test of Haskins 'sexual selection' predation hypothesis. *Environmental Biology of Fishes* **29**, 179–191.

Winge, Ø. 1922*a*. One-sided masculine and sex-linked inheritance in *Lebistes reticulatus*. *Journal of Genetics* **12**, 145–162.

Winge, Ø. 1922*b*. A peculiar mode of inheritance and its cytological explanation. *Journal of Genetics* **12**, 137–144.

Winge, Ø. 1927. The location of 18 genes in *Lebistes reticulatus*. *Journal of Genetics* **18**, 1–43.

Winge, Ø. 1937. Succession of broods in *Lebistes*. *Nature* **140**, 467.

Winge, Ø. and Ditlevsen, E. 1947. Colour inheritance and sex determination in Lebistes. *Heredity* **1**, 65–83.

Winge, Ø. and Ditlevsen, E. 1948. Colour inheritance and sex determination in *Lebistes*. *Comptes rendus des travaux du. Labratoire. Carlsberg, Ser. Physiol.* **24**, 227.

Wootton, J. T. 1993. Indirect effects and habitat use in an intertidal community: interaction chains and interaction modifications. *American Naturalist* **141**, 71–89.

Wootton, J. T. 1994. The nature and consequences of indirect effects in ecological communities. *Annual Review of Ecology and Systematics* **25**, 443–466.

Wootton, R. J. 1990. *Ecology of teleost fishes*. London: Chapman and Hall.

Wourms, J. P. 1981. Viviparity: the maternal-fetal relationship in fishes. *American Zoologist* **21**, 473–515.

Wright, D., Rimmer, L. B., Pritchard, V. L., Krause, J., and Butlin, R. K. 2003. Inter and intra-population variation in shoaling and boldness in the zebrafish (*Danio rerio*). *Naturwissenschaften* **90**, 374–377.

Wright, P. J. and Huntingford, F. A. 1992. Inherited population differences in avoidance learning in three-spined sticklebacks (*Gasterosteus aculeatus*). *Behaviour* **122**, 154–174.

Wu Chingjiang and Schröder, J. H. 1984. Monomorphic and polymorphic isozymes in laboratory strains of guppies (*Poecilia reticulata*) Peters. *Biologisches Zentrablott* **10**, 61–67.

Yamagishi, H. T. O., Nakamoto, N., Nakamura, Y., and Wade, Y. 1967. Ecological studies on the guppy *Lebistes reticulatus* (Peters). III. On the guppy population acclimatized in a water of Togura-Kamiyamada spa. *Japanese Journal of Ecology* **17**, 206–213.

Zambrano, L. and Macías-Garcia, C. 1999. Impact of introduced fish for aquaculture in Mexican freshwater systems. In *Nonindigenous freshwater organisms: Vectors, biology and imacts* (ed. R. Claudi and J. H. Leach), pp. 113–124: Lewis Publishers, Boca Raton, FL, USA.

Zhuikov, A. Y. 1993. Avoidance learning and aggression in guppies. *Animal Behaviour* **45**, 825–826.

Index

Note: there are no entries under 'guppies' or '*Poecilia reticulata*', as these constitute the subject of the entire book. The reader is advised to search under a more specific heading.